THE GRENVILLE PROBLEM

The Grenville Problem

THE ROYAL SOCIETY OF CANADA
SPECIAL PUBLICATIONS, NO. 1

Edited by James E. Thomson

PUBLISHED BY THE UNIVERSITY OF TORONTO PRESS

IN CO-OPERATION WITH

THE ROYAL SOCIETY OF CANADA

1956

SCHOLARLY REPRINT SERIES
Reprinted in 2018
ISBN 0-8020-7022-1
ISBN 978-1-4875-7348-5 (paper)
LC 57-32022

PREFACE

THE GRENVILLE SUBPROVINCE forms an important part of the Precambrian Shield in eastern Canada. It covers an area 150 to 200 miles wide extending along the southern border of the Shield from Lake Huron to the Labrador coast and projects into the Adirondack region of the United States. It has a total area of about 250,000 square miles of which about 10,000 square miles are in New York State, 25,000 square miles in Ontario, 177,000 square miles in Quebec, and 38,000 square miles in Labrador. This vast expanse of territory is distinguished from the remainder of the Shield by the presence of extensive areas of crystalline limestone intermingled with a variety of gneisses and highly metamorphosed sedimentary and volcanic rocks. Because of the great geological complexity and rock alteration the Grenville subprovince presents a variety of problems that have always puzzled and intrigued geologists. A century of investigation has produced much factual information about the distribution of rock types, their relationships, alterations, structure, age, and valuable mineral content. This in turn has led to much speculation as to their mode of origin and relationship to adjacent Precambrian units. Although the Grenville subprovince has not been a prolific source of mineral production when compared with some other major subdivisions of the Shield, it contains the greatest variety of mineral species, and, in recent years, important deposits of iron ore, lead and zinc, uranium, and industrial minerals have been developed. This encouragement has greatly increased the incentive for mineral exploration throughout the whole region. The Grenville subprovince is now undergoing the greatest period of mineral search and development in its entire history.

With these facts in mind, the officers of Section IV (Geology and Allied Sciences) of the Royal Society of Canada decided that it was an opportune time to discuss all aspects of Grenville geology. Accordingly, a symposium on "The Grenville Problem" was arranged for the annual meeting of the Society at Toronto in June, 1955. Geologists with considerable experience in all aspects of Grenville geology were invited to contribute papers and take part in the discussions. They represented mining companies, government surveys, and universities. Thirteen papers were presented and discussed as fully as time would permit. It was felt that this information should be made available to a wider audience, and eventually, through the joint efforts of the Royal Society of Canada and the University of Toronto Press, several of the papers presented on that occasion were assembled in this publication.

The symposium does not attempt any systematic coverage of the Grenville rocks nor does it contain contributions from all geologists who have specialized in Grenville geology. Rather, it is a random sampling of areas and topics that are of current interest. It discusses a wide variety of subjects and questions that have come to the fore as geologists delve deeper into the mysteries of the Grenville rocks. They ask: What is the precise meaning of "Grenville" in so far as the term applies to rock types, localities, and concepts? What are the main local units and how are they recognized and differentiated? What is the exact age of the Grenville rock complex? What is its relationship to adjacent subprovinces? What about granitization, intrusion, structure, orogeny? How is the "Grenville front" to be explained? The authors struggle with these and many other problems, often with reference to a specific locality but in some cases dealing with the Grenville as a geological unit. It is not to be expected that students of the Grenville rocks will agree with all that is presented here, but they will be stimulated and impressed by the amount of useful information that patient and systematic field and laboratory work can dig out of a seemingly hopeless rock complex.

It is a particular privilege to have papers from two American geologists, Professor A. F. Buddington and Professor A. E. Engel. They have had much experience in the Adirondack region of New York State and are keenly interested in Grenville problems. Their contribution adds much to the value of the symposium.

This is the first Canadian attempt to collect some of the available information and ideas on a difficult and fascinating subject. That it is incomplete and has several omissions is admitted. It is hoped, however, that it will serve a useful purpose and inspire further study and discussions on a geological unit that has a very considerable future potential in the Canadian mineral economy.

JAMES E. THOMSON,
Chairman, Editorial Committee

CONTENTS

CONTRIBUTORS

JAMES E. THOMSON (editor), Ontario Department of Mines.

F. FITZ OSBORNE, Faculté des Sciences, L'Université Laval, Québec.

W. G. ROBINSON, Frobisher Ltd., Noranda, P.Q.

D. F. HEWITT, Ontario Department of Mines.

H. A. SHILLIBEER, Gulf Research and Development Corp., Pittsburg, Pa., formerly of the Geophysical Laboratory, University of Toronto.

G. L. CUMMING, British American Oil Company, Calgary, Alta., formerly of the Geophysical Laboratory, University of Toronto.

A. E. ENGEL, California Institute of Technology, Pasadena, Calif.

J. W. AMBROSE, Queen's University, Kingston, Ont.

A. F. BUDDINGTON, Princeton University, Princeton, N.J.

C. A. BURNS, W. C. Ringsleben & C. A. Burns, Consulting Mining Geologists, Toronto.

THE GRENVILLE PROBLEM

THE GRENVILLE REGION OF QUEBEC*

F. Fitz Osborne, F.R.S.C.

"GRENVILLE" AND "LAURENTIAN" as names for rocks came into use in Quebec when Precambrian rocks were considered so different from fossiliferous rocks that the application to them of the ordinary rules of stratigraphy was doubtful. In those happier times, stratigraphic terms even for fossiliferous rocks were used less rigorously than now. It has thus been possible for a term such as Grenville series, which was very casually and loosely defined, to be useful to three generations of geologists. This is in spite of the fact that diversities within the Grenville in Quebec have long been recognized and emphasized, notably by R. W. Ells. Rocks in southeastern Ontario, the Adirondacks of New York, and elsewhere have been called Grenville, although in many respects these rocks differ from those of the type locality in Quebec. The Grenville rocks outside Quebec are in small and generally homogeneous units so that there has been a tendency to consider that some of their dominant features are applicable to all Grenville. An urge to re-define Grenville has thus arisen. It is a matter of opinion whether it is desirable to do this with the knowledge of the Grenville now available. Samuel Butler said, "Definitions are a kind of scratching, and generally leave a sore place more sore than it was before." What must re-definition be? Many changes in definition had unhappy results. The effect on Grenville of the change in Laurentian is outlined in the historical section of this paper. Logan's Quebec group is another example: elimination of this group caused Sillery to assume some of its function, and as a result Sillery has become a practically useless name.

I contend that Grenville series has a place in geological nomenclature, and the place is justified by history and usage, although it is unlikely that the same name would be proposed now.

Most of the significant history of Laurentian and Grenville as geological names can be summarized as pertaining to Quebec only. Furthermore, the Grenville rocks are considered to be south of latitude 50°. This is a matter of convenience only.

Grenville Geology before Logan

When the geology of Quebec or Ontario is discussed, it is habit to "go back to Logan." Such a tribute to Logan's genius tends to obscure the fact

*Published with the permission of the Deputy Minister of Mines, Quebec.

3

that some geological observations were made prior to Logan's appointment in 1842 as Director of the Geological Survey of Canada. The differences between the sand or clay mantled lowlands along the St. Lawrence River and the rocky uplands of the Laurentide mountains were obvious to the earliest European settlers. Some of the characteristics of bedrock were noted by explorers, although observations were casual until after the beginning of the nineteenth century.

Three early geologists were Baddeley, Bayfield, and Bigsby. Of these Baddeley is my favourite. In 1828 he was attached to a party which explored part of the north shore of the St. Lawrence River, the Saguenay River, and Lake St. John and its vicinity. His account of the geognosy shows that he realized many of the essential problems of the rocks. He was concerned over the origin of the dark (amphibolite) inclusions in the gneisses and whether the layers in the granite (biotitic) gneisses and the syenite (hornblendic) gneisses were relic beds or secondary structures. He mentioned the occurrence of "primitive limestone," although the occurrence at Moulin Baude brook, to which he devoted the longest description, is now known to be a vein. He considered the anorthosite of Lake St. John a variety of syenite. It is exaggerating only slightly to say that Baddeley's work shows his understanding of Laurentian problems to be superior to that achieved by Logan.

Laurentian System of Logan

Logan had the greatest responsibility for the terms Laurentian and Grenville. Although he was primarily a stratigrapher, and the crystalline rocks of the Laurentian uplands were probably less interesting than the fossiliferous formations to him, he made a traverse, in 1845, along the Ottawa River from Mattawa to the north end of Témiscamingue Lake. He thus crossed to the northwestern limit of the Grenville subprovince. He (p. 40) inferred that the major structure there is anticlinal and described the rocks as follows:

The lowest rocks which this undulation brings to the surface are of a highly crystalline quality, belonging to the order which, in the nomenclature of Lyell, is called metamorphic instead of primary, as possessing an aspect inducing a theoretic belief that they may be ancient sedimentary formations in an altered condition. Their general character is that of syenite gneiss. Their general colour is reddish and it arises from the presence of reddish feldspar, which is the prevailing constituent mineral.

This description refers in large part to rocks that were later termed "Ottawa gneiss." Logan believed that the limestones were less abundant in the lower part of the section. In the same report, he stated (p. 50): "From the vicinity of Quebec, the formation to which the metamorphic limestone belongs ranges along the St. Lawrence, at a distance from its margin varying from ten to twenty miles." He also mentions the occurrence of crystalline limestone near Grenville, as the first to reach the shore of the Ottawa River.

It became evident that the name "metamorphic" was not sufficiently distinctive for the crystalline rocks, and Logan, therefore, proposed "Laurentian series" in the report for 1852–53 (dated 1853, published 1854). In the outline of the geology of Canada by Logan and Hunt, published in Paris in 1855, *système laurentien* is used despite the fact that *formation laurentienne* also appears, with the same meaning. In *Geology of Canada*, 1863, "Laurentian system" is used in chapter III, and "Laurentian series" appears in chapter XXII, the one in which "Grenville series" is mentioned. In several publications, Hunt states that he was co-proposer of "Laurentian," and finally in 1888 he wrote of Laurentian, ". . . sous ce nom proposé et adopté par l'auteur en 1854."

Despite some of the uncertainties regarding the term Laurentian, there is no doubt that Logan intended the name to apply to metasedimentary rocks. The stratigraphy was a challenge to him, and he started field work in Argenteuil county to establish it. This became the "original Laurentian area" and received much attention not only from Logan but also from his co-workers. Map 12, dated 1856, shows the "distribution of limestones of the Laurentian series" between Grenville and Rawdon (it is of interest as the first coloured map issued to illustrate a report of the Geological Survey). Hunt, who joined the staff of the Survey while the work was in progress on the original Laurentian area, was a good mineralogist, but unfortunately Logan accepted his views on the genesis of crystalline rocks. Hunt was the last nineteenth-century Wernerite, and he considered all crystalline rocks to be of aqueous origin. Logan apparently had some reservations and in his reports mentions the resemblance of some facies of the gneisses to eruptive rocks. In the original Laurentian area, the intrusives are small and conformable so that a sequence which appeared reasonable was built up for the Laurentian.

As work was extended eastward, the Morin anorthosite massif was encountered. Anorthosite had been met before; for example, the body near Baie St. Paul had been considered a syenite facies of the "Metamorphic Group" by Logan (1849–50, p. 8), who thus followed Baddeley's determination. The peculiar mineralogy of the anorthosite was recognized by 1852 (Logan, 1853–56, p. 35), and "lime-feldspar rock," "andesine, or labradorite rock," became distinctive members of the Laurentian system. The intrusive nature of anorthosites was recognized outside Canada, but Logan, apparently at Hunt's instigation, considered the rocks metasedimentary, and because, in places, the structure of the anorthosites is athwart the structure of the gneisses, the anorthosites together with some gneisses and limestone were assigned an upper unconformable position in the Laurentian system.

The Upper Laurentian and the anorthosite were known by various names: labradorite gneiss, anorthosite gneiss, Labrador series, Labradorian. Hunt proposed norite for the rocks in Canada, and in 1871, he proposed Norian as a synonym for Upper Laurentian.

Logan's Grenville Series

During the working out of the stratigraphy of Laurentian system, the several bands of limestone received names such as Trembling Mountain Lake, Green Lake, Morin, and Grenville. The term "Grenville series" was first used in chapter XXII (p. 839) of *Geology of Canada*, 1863. It is appropriate to quote the first reference. "From the interruption by it of the Morin limestone near Howard, it seems probable that the anorthosite rock overlies the whole Grenville series unconformably, and that the mass of it on the west side of Desalaberry is an outlying portion." It is difficult to believe that this was intended as a formal proposal of a series name. The 1863 report has marginal notes, and no note appears opposite the first mention of Grenville series. Furthermore, in the report for 1863–66 neither Logan nor Macfarlane used Grenville series. Logan mentions Grenville "band" of "Laurentian limestone." It may be noted that in some publications, Hunt claims co-authorship of "Grenville series."

In 1878 Hunt (p. 155E) proposed that "Ottawa gneiss" be used for the orthoclase gneiss presumed to underlie the Grenville series. These rocks had been referred to by Logan as "fundamental gneiss" but more commonly as part of the Lower Laurentian. Barlow (1897, p. 531) used "Ottawa series" for the same rocks.

At the time of Logan's death in 1875, it must have appeared that the Laurentian was a well-established system. However, considerable doubts about the validity of including the anorthosites in a presumed metasedimentary sequence had arisen. Vennor (1876–77, p. 254) had such doubts, and finally Selwyn (1877–78, p. 12A), Logan's successor as director of the Geological Survey, wrote:

If it is admitted—which, in view of the usual associations of Labrador felspars, is the most probable supposition—that these anorthosite rocks represent the volcanic and intrusive rocks of the Laurentian period, then also their often massive and irregular, and sometimes bedded character, and their occasionally interrupting and cutting off some of the limestone bands, as described by Sir W. E. Logan, is readily understood by anyone who has studied the stratigraphical relations of contemporaneous volcanic and sedimentary strata of palaeozoic, mesozoic, tertiary and recent periods.

Decline of Laurentian and Rise of Grenville

The Laurentian system seems to have become less significant with its decapitation and with the realization that many of the gneisses were of igneous origin. The orthogneisses of the Laurentian region, where they were not intimately associated with para-rocks, were inferred to be of igneous origin, although it was fashionable to regard them as palingenic products from an original crust so that the name "Fundamental gneiss" was retained for them. Ells (1897, p. 124) proposed: "Laurentian, non-sedimentary, Basal or Fundamental Gneiss (Ottawa gneiss) representing in altered form the original crust of the earth, and the lowest known series of rocks; without evidence of sedimentary origin." Grenville and Vennor's Hastings series were relegated to the Huronian, which at that time included the Keewatin.

The destruction of the Laurentian system was completed when an international committee (Adams *et al.*, 1907, p. 216) recommended that Laurentian be restricted to orthogneisses intrusive into the Grenville. This usage was in assumed conformity with the usage recommended by a committee for the correlation of rocks near Lake Superior. The name Laurentian was allocated to intrusive rocks older than Huronian. The decisions of the committees were apparently not very happily received in Canada. Adams presented a paper entitled "The Laurentian System in Eastern Canada" in 1908, despite the fact that he signed the reports of both committees.

It was a long time before "Grenville series" was generally used, and it became a common name only after the misappropriation of Laurentian. Indeed it had a tendency to assume some of the duties done by Laurentian. From indices to the publications of the Geological Survey an estimate of the use of Grenville can be made. The index for 1863 to 1884 has only three entries under "Grenville series," and one is a reference to the proposal of the name. In the volume for 1885 to 1906, there are still not many entries as compared with those under "Grenville township." In the 1905–16 volume, "Grenville series" entries far exceed those under "Grenville township." If more evidence of the slow acceptance of the name is required, it is to be found in the way in which the series is referred to in some publications from 1890 to 1895. Grenville series is enclosed within quotation marks: "so-called Grenville series"; "Logan's Grenville series."

At present, Grenville series is a relic of the Laurentian system. Fundamental gneiss and Ottawa gneiss are not generally current, and Norian has been shown to be improperly applied. The geology of the Laurentian region is complex and Grenville series has a heavy load to bear. Furthermore, the name has been extended to such unrelated and perhaps hypothetical things as "Grenville front" and "Grenville orogenic belt."

INVESTIGATION OF GRENVILLE ROCKS

Geological Survey of Canada

In the preceding section of this paper, the history of "Laurentian" and "Grenville" as geological terms has been reviewed. From it, it is easy to see how scanty are the data for defining "Grenville series." Actually the term has not been static but has developed with use, and the use of the term has been in a considerable measure based on the areas of the Laurentian region examined. After the time of Logan, some maps were prepared by Vennor, who principally followed the methods of Logan. Ells is the first who mapped according to reasonable geological standard large areas in the Grenville region. He is responsible for the geology of a series of maps on a scale of one inch to four miles. Ells has mapped a greater area of the Grenville subprovince than any other geologist, and one can also get a better appreciation of the characteristics of the Grenville series over the large area from his than from any other reports. In some of his reports he does not use the term Grenville series. In places, he expresses doubts about the certainty of correlation of the paragneisses with the Grenville series of the type locality. Ells was primarily a field geologist and his descriptions of rocks, particularly igneous rocks, are scanty. These deficiencies were supplied by the work of Adams, Barlow, and Ferrier. Adams's (1895 J) report on the region north of Montreal contains petrographic descriptions and analyses of both orthogneiss and paragneiss. The region is east of the type locality, having been mapped largely as Upper Laurentian, and Adams thus concentrated most of his work on rocks unlike the type Grenville and on the orthogneisses. Adams was prone to publish the same data in many papers and as a result his conclusions gained greater currency than those of Ells did, despite the fact that Ells was the superior field geologist. Furthermore, the thick limestones of the Haliburton and Bancroft areas were repeatedly described by Adams, and thus he seems to have done much not only to establish the abundance of orthogneisses but also to establish the Grenville as a limestone sequence.

From 1910 to 1930, mapping at one inch to a mile was done for the Geological Survey of Canada largely by M. E. Wilson, but it was almost restricted to the original Laurentian region or its extension. This was partly the result of the fact that the original Laurentian has deposits of industrial minerals. The mapping tended to confirm the notion that the Grenville series has much limestone.

Quebec Department of Mines

Mapping under the direction of J. A. Dresser by the Quebec Department of Mines started in 1927. A series of traverses searching for limestone in regions between the Gatineau and Saguenay rivers showed the erratic distribution of limestone, and the system of traverses was abandoned. Since 1930 most of the mapping in the Quebec Grenville subprovince has been by geologists of the Quebec Department of Mines. Map sheets in the original Laurentian area and its extension show abundant carbonate rocks, whereas map sheets outside this zone show little or, in places, no carbonate rocks. A convention grew up that any obviously metamorphic rocks, probably originally sedimentary or volcanic, are referred to the Grenville series. More recently, "Grenville-type" has come into use. The meaning of this term is not clear in all cases. It obviously should refer to paragneisses, but it has been used by some for orthogneisses.

CRITERIA FOR GRENVILLE SERIES

Certain criteria for Grenville series have been used, if not implicitly at least inferentially, and the validity of some of these must be examined.

Crystalline Limestones

In any discussion of the Grenville series, limestones are likely to be mentioned prominently. Logan placed great emphasis on them, partly because they were useful as land lime and hence tracing them was part of the justification for the expense of the Geological Survey, and partly also, one suspects, because of their distinctive characteristics. As mentioned above, the type area for both Grenville and Laurentian is a carbonate-rich province: carbonate and the derivative rocks certainly are in excess of 15 per cent of the paragneisses, whereas northeast of the type locality, limestones do not amount to 1 per cent of the paragneisses. There has been a tendency to emphasize the presence of carbonate and to interpret any carbonate as limestone. Some reported occurrences of limestone are vein carbonate. Then, too, a thin layer of limestone has been considered enough evidence to assign some paragneisses to the Grenville series. Such a procedure might be valid if Grenville were the only series known to have any carbonate. However, some carbonate-bearing rocks have been reported in the Timiskaming sub-province. Cooke (1919, p. 373) has described two beds of limestone, each a foot thick, from Nemenjish series. Bain (1925, p. 739) described limestones from "pre-Keewatin sediments" from the basin of the Harricana River, and later Tolman (1932) described limestones from "an early pre-Cambrian sedimentary series," in Quebec. Cooke noted that the intensity of metamorphism of the Nemenjish series increases towards the Grenville sub-province, and newer mapping shows that these rocks are involved in zones of high grade metamorphism. In addition, the Pontiac series has a composition such that limestones can be expected. Certain silicate layers in it were probably impure limestones. It is indeed probable that the limestones reported from the Grenville subprovince southeast of Senneterre are in the Pontiac series. In any case, several Archean sedimentary series are known to have limestones, and, hence, limestones are not exclusively Grenville.

Any definition involving limestone would necessarily be based on the proportion of limestone among the paragneisses. The setting of a precise ratio then becomes difficult, for several reasons, the principal one being that limestone can be converted to non-carbonate rock, although this reason is, in my opinion, not serious.

Using a definition for Grenville based on 2 per cent or more limestone in the paragneisses cropping out in an area of reasonable size and with a high grade of metamorphism, one gets closest to the condition of the type locality for the Middle Laurentian or the upper part of the Lower Laurentian. Defined in this way, the Grenville series in Quebec crops out for 100 miles west of the type locality and about 100 miles northward from the Ottawa River. This is an area of 10,000 square miles. An insignificant north-striking belt extending north of Lake St. John could also be considered Grenville.

Quartzite

The quartzites are irregularly distributed. Thin interbeds with limestone or sillimanite-garnet gneiss are common. Thick massive pure quartzite is rarer and where it occurs, suggests it may be near the base of the Grenville as defined above. The quartzite of the Wakeham Lake beds is exceptional, but it is north of latitude 50°. Many gneisses have about 50 per cent quartz and may well have been dirty sandstones. It is noteworthy that these rocks are not common with limestone.

Hornblende Gneiss and Amphibolite

Gneissic or massive metamorphic rocks consisting of plagioclase and amphibole and at some localities pyroxene have been assigned to the Grenville in many areas. The reason for the assignment is not clear, but perhaps it stems from the notion that the gneisses or amphibolites of this type are derived from limestones. It is certain that rocks of this composition have formed from basic dykes, sills, flows, tuffs, and, in at least one locality, from anorthosite.

The distribution of these rocks is not uniform. They occur but are not abundant in the region north of the Ottawa River. Ells, in discussing the region in Ontario south of the Ottawa River (Adams, Barlow, Ells, 1897, p. 180), has commented on the distribution: "The areas of limestone became much more extensive, and there was a large development of hornblende and other dark-colored rocks, rarely seen to the north of the Ottawa. The limestones also were very often highly dolomitic, and in certain areas were blue and slaty, with but little of the aspect of the Grenville limestones." However, east of the type locality the hornblende paragneisses are again common. Rocks of this composition are particularly abundant southeast of the hypothetical "Grenville front." Such rocks can be traced northwestward into pillowed greenstones and "older gabbro." They are then the result of a high intensity of metamorphism and are thus useful criteria for Grenville only to the extent that high grade metamorphism is an attribute of the Grenville subprovince.

Garnet

A Wernerian touch is given to the problem of Grenville correlation in the significance attached to garnet. Despite the fact that in many parts of the Grenville subprovince garnets are scarce or even absent, a tendency has grown up to consider garnet diagnostic of the Grenville series. A quotation from Cooke (1919, p. 371) is cogent in this connection. "Many descriptions were met with of rocks which bear a strong resemblance to certain phases of the Grenville, but are likewise indistinguishable from the Pontiac or some other highly altered sedimentary series. These are not included on the accompanying map; the bodies mapped are confined to those highly garnetiferous types, with or without limestone, which are peculiarly characteristic of the Grenville." It is now known that garnet is a mineral of varied provenance and may form under many and diverse conditions, so that it is useless as an index mineral for Grenville series.

Environment of Sedimentation

Another possible method of defining the characteristics of the Grenville series is to identify it as a sequence of rocks characteristic of one type of sedimentary environment. This possibility has been little exploited directly, but in discussions of correlatives of the Grenville series it has been obvious that if the Grenville has correlatives among rocks close to it, a pronounced change in facies must be present. Almost every possibility has been suggested. For example, the Keewatin rocks have been considered terrigenous, and the Grenville has been considered an off-shore facies of the Keewatin. The Grenville has been considered an off-shore facies of the Timiskaming. More recently, the interpretations have been reversed and the Grenville series has been inferred to be the shelf facies with the Keewatin or Timiskaming series as the deep geosynclinal facies.

This is a promising method of approach to the problem of the origin of the Grenville series, but its application is perhaps better deferred until the criteria for environments have been better established.

Relationship of Folding and of Igneous Intrusions

In the Grenville region in Quebec, the paragneisses are metamorphosed to a high grade, the associated igneous rocks penetrate them intimately, and many of the masses are conformable. Many of the igneous rocks are gneissic[1] so that structures parallel to the metasediments can be mapped in them. This situation is in contrast to that found in the metavolcanic rocks in the Val d'Or–Noranda belt of the Timiskaming subprovince, where a greenstone grade of alteration is characteristic and the intrusives are principally transgressive. The lithologies are sufficiently contrasted that the strongly altered gneisses have been considered Grenville, although it has been demonstrated in several areas, since 1936, that there is a gradation in crossing the boundary between the Grenville and Timiskaming subprovinces. This relationship invalidates alteration as a criterion for Grenville although, along with the presence of garnet, it has been one of the criteria most used.

In the original Laurentian region and for a considerable distance west and north of it the dips are steep and the folding tight. The para-rocks have been very plastic during deformation, and it has proved difficult to trace structures for long distances along strike. Only a few folds have been traced as much as fifteen miles and most have not been traced more than three miles. Strikes are moderately consistent, commonly north. The intrusives are small, although abundant. In much of the Grenville subprovince, however, the dips are not steep, but broad domical structures are found. The intrusives are more abundant than in the original Laurentian area and more of them are transgressive. The degree of metamorphism and the general aspect of the intrusives is the same in both structural regions.

Anorthosites have been considered peculiarly characteristic of the Grenville subprovince, perhaps because of the assumed stratigraphic position of the Norian in the Laurentian system. However, except insofar as anorthosites may be an index of the physical conditions of an environment, the anorthosites are without value in correlation. The best known anorthosites

[1] A deplorable tendency has grown up to refer to some coarsely crystalline orthogneisses as Grenville, without mention of the fact that the rocks are orthogneisses.

of the Grenville region, such as the massifs of Morin and Saguenay, have a domical habit. Stratiform masses, like the Bell River, Opawica, and Chibougamau anorthosites, do not occur in the Grenville subprovince although they may be represented by layered metatroctolite and coronite.[2] Anorthosite replacing paragneiss and possibly limestone forms small bodies in the Grenville subprovince.

The several modes of occurrence of anorthosite make it an unreliable criterion for Grenville. Furthermore, the larger massifs of anorthosite are not obviously related to outcrops of the typical Grenville rocks. This fact is probably the reason for the differences of opinion regarding the relationship of the domical anorthosites to intrusion and folding. It is certain that on the north shore of the St. Lawrence River, paragneisses were folded and intruded by granitic rocks before the intrusion of the anorthosite. The relationships are more equivocal near the Morin massif.

CONCLUSIONS

The summary of the history of the use of Grenville and its characteristics suggest to me that the name "Grenville series" be retained. It should be a matter of indifference that some geologists consider that series suggests a time stratigraphic unit, for as yet there is no proof that the Grenville is not such a unit. An Archaean age for the Grenville has not been disproved. Within the Grenville subprovince of Quebec several different kinds of units may be recognized. At the type locality and for a region about it, a carbonate-bearing type of Grenville is common. At other localities, metaclastic rocks without carbonate are found, and finally, in extensive areas, hornblende plagioclase gneisses perhaps derivative from volcanics are dominant. The two latter groups may well be altered equivalents of metavolcanic and metasedimentary rocks of the Timiskaming subprovince. Some evidence suggests that the carbonate Grenville is high in the series and is possibly separated from the clastic members by an unconformity, but the areal mapping has not advanced to such a stage that this can be proved.

If the definition of Grenville series is lacking in precision it is in this similar to other definitions for Precambrian rocks. Fundamentally most such definitions are conventions. Thus the term "Keewatin" suggests an old predominately metavolcanic series. There is no guarantee that all rocks called by this name are the same age or were formed in a similar environ-

[2]Cooke (1919, p. 274) offered the suggestion that the anorthosites are of one age. Thus the layered anorthosite complexes of the Timiskaming subprovince would be coeval with the domical masses of the Laurentian region. Furthermore, he presented evidence, if not proof, that Mattagami rocks, later correlated with Timiskaming, are post-anorthosite. The presence of anorthosite boulders in conglomerate involved with high grade gneisses with a northeast strike was confirmed by a field party of the Quebec Department of Mines in 1953. In view of the fact that the domical anorthosites cut Grenville rocks folded and intruded by granites, Cooke's hypothesis would make the Grenville pre-Timiskaming. This seems unlikely because the north-south striking Grenville probably cuts across east-west striking correlatives of Keewatin and Timiskaming. The northeast trending "Grenville orogenic belt" appears to cut across both east-west and north-south folding.

ment. The problems of the Timiskaming-type rocks are fully as vexatious as those of the Grenville. Grenville is acceptable in a stratigraphic sense. The only question is whether it should be restricted to one lithology, such as that of the original Laurentian area. If this were done some name would be necessary to do that work now done by Grenville. "Laurentian" is an obvious choice. Should the name be revived?

While I was writing this, a thought kept recurring: "Grenville is neither a series nor a lithological unit. It is a state of mind." Of course, rocks do not have a state of mind, but they can engender one in those who work with them. I feel that progress has been made and that the knowledge of the geology of Grenville subprovince is a matter of pride. The excessively critical and defeatist approach to Grenville problems is not justified.

REFERENCES

ADAMS, F. D. (1896). Geology of a portion of the Laurentian area lying to the north of the island of Montreal; Geol. Surv., Canada, Ann. Rept., vol. 8.
———— (1908). The Laurentian system in eastern Canada; Quart. Jour. Geol. Soc., vol. 64, pp. 127–143.
ADAMS, F. D. and BARLOW, A. E. (1897). On the origin and relations of the Grenville and Hastings series in the Canadian Laurentian; Am. Jour. Sci., 4th Series, vol. 3, pp. 173–180.
ADAMS, F. D., et al. (1907). Report of a special committee on the correlation of the Precambrian rocks of the Adirondack mountains, the "original Laurentian area" of Canada and eastern Ontario; Jour. Geol., vol. 15, pp. 191–217.
BADDELEY, F. H. (1829). Geognostical section through part of the Saguenay country; Report of the commissioners for exploring the Saguenay.
BAIN, G. W. (1925). Pre-Keewatin sediments of the upper Harricana basin, Quebec; Jour. Geol., vol. 33, pp. 728–743.
BARLOW, A. E. (1897). Report on the geology and natural resources of the area included by the Nipissing and Temiscaming map-sheets; Geol. Surv., Canada, Ann. Rept. 1897, vol. 10, pt. I.
COOKE, H. C. (1919). Some stratigraphic and structural features of the Pre-Cambrian of northern Quebec; Jour. Geol., vol. 27.
———— (1920). A correlation of the Pre-Cambrian formations of northern Ontario and Quebec; Jour. Geol., vol. 28, p. 304–332.
ELLS, R. W. (1897). Notes on the Archaean of eastern Canada; Trans. Royal Soc. Can., Sec. IV, pp. 117–124.
HUNT, T. S. (1878). Special report on the trap dykes and azoic rocks of southeastern Pennsylvania; Pt. I, Second Geological Survey of Pennsylvania.
———— (1888). Les schistes crystallins; C. R. Int. Geol. Cong.
LOGAN, W. E. (1847). Report of Progress for the year 1845–46; Geol. Surv., Canada.
———— (1850). Report of Progress for the year 1849–50; Geol. Surv., Canada.
———— (1854). Report of Progress for the year 1852–53; Geol. Surv., Canada.
———— (1857). Report of Progress for the years 1853–54–55–56; Geol. Surv., Canada.
———— (1863). Geology of Canada; Geological Survey of Canada, Report of Progress from its commencement to 1863.
LOGAN, W. E. and HUNT, T. S. (1855). Esquisse géologique sur le Canada. Paris, 1855.
SELWYN, A. R. C. (1877–78). Report of observations on the stratigraphy of the Quebec group and the older crystalline rocks of Canada; Geol. Surv., Canada, Rept. Progress, 1877–78, pt. A.
TOLMAN, CARL (1932). An early pre-Cambrian sedimentary series in northern Quebec; Jour. Geol., vol. 40, pp. 353–373.
VENNOR, H. G. (1876–77). Counties of Renfrew, Pontiac and Ottawa; Geol. Surv. Can., Rept. Progress, 1876–77.

THE GRENVILLE OF NEW QUEBEC

W. G. Robinson

THIS PAPER DEALS WITH the easterly extension of the Grenville geological province into Ungava and Labrador. For three years, Frobisher Limited had a 3500 square mile mining concession centred around Seal Lake in Labrador. The concession was mapped by the Geological Survey of Canada (7, 14), and by Frobisher geologists. The Seal Lake area is roughly in line with the projected extension of the northern edge of the Grenville geological province, and a description of the geology may supply a few clues to aid in the solution of "the mystery of the Grenville." Information from other sources, pertaining to the easterly extension of the Grenville, will also be briefly reviewed.

GEOLOGY OF THE SEAL LAKE AREA

The Seal Lake concession is underlain by intrusives, sedimentary formations, and volcanics of Pre-cambrian age. These can be grouped into five major divisions. In the northwest part is the flat-lying Shipiscan series; in the eastern part is the folded Croteau Lake series; in the central part is the folded and faulted Seal Lake series; south of the Seal Lake series is the Bessie Lake series; and south of the Bessie Lake series are granites, gneisses, and basic intrusives.

The Shipiscan series consists principally of flat-lying and gently dipping basic flows, sandstones, and red argillites, which were deposited on eroded granites and anorthosites (10). The formations become folded near the southern boundary with the Seal Lake series, and in one place appear to be intruded by a granite boss. The Shipiscan series is probably about 1500 feet thick.

The Croteau Lake series consists of porphyritic flows of intermediate to acidic composition, basic flows, sandstones and quartzites, cherts, conglomerates, black pyritic slates, and minor beds of limestone. In the northeast part of the concession, these formations form an open anticlinal fold, and a tongue of grey granite intrudes along the northeasterly trending fold axis. On the north flank the formations dip about 50 degrees to the northwest, and on the south they dip about 50 degrees south. The series is bounded by younger granites to the north and south. It fingers out in the granites to the east (17) but may be equivalent to areas of altered sedimentary rocks near Cape Makkovic (13). To the west, the Croteau Lake series is overlain by the Seal Lake series thrust fault block.

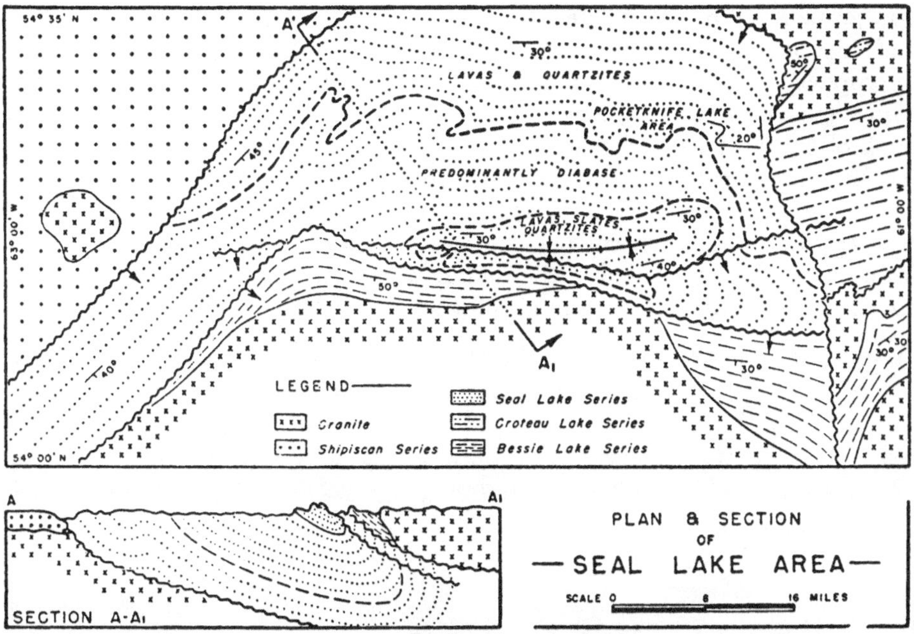

FIGURE 1

The Seal Lake series consists of sedimentary and volcanic formations that have been folded, faulted, and intruded by diabase sills. The series has an apparent maximum thickness of 50,000 feet, but this has probably been exaggerated by faulting, folding, and intrusion of sills. The formations form a huge arc, convex to the north. They have been folded into a major synclinal structure which has been overturned with nearly all of the members dipping south at angles of 30 to 60 degrees. The series is interpreted as being part of a fault block, thrust from the south, which overrides the Shipiscan series, granites, and anorthosites on the north, and the Croteau Lake series on the east. At Pocketknife Lake the Seal Lake formations strike northerly and dip 15 degrees west, and the adjacent Croteau Lake formations strike westerly and dip steeply south. The sedimentary and volcanic members of the Seal Lake series are lithologically similar to those of the Shipiscan series, and it is probable that the Seal Lake series is the deformed equivalent of the Shipiscan. The geological sequence of the Seal Lake series is:

Upper red quartzites

Interbedded lavas and red slates

Red and grey argillites with interbedded quartzites and some diabase sills

Diabase sills with slates, argillites and quartzites

Green lavas with quartzites and thin conglomerate beds

Basal conglomerate (indicated by float)

The Bessie Lake series consists of grey quartzites and basic flows that show an increasing grade of metamorphism along the southern contact and appear to have been intruded by granites. The formations strike easterly and dip south. Together with adjacent granites and gneisses, they are interpreted as being part of another fault block, thrust from the south, which overrides the Seal Lake fault block and covers much of the south flank of the Seal Lake synclinal structure. The Bessie Lake series could be the faulted equivalent of the lower members of the Croteau Lake series.

Thus the Seal Lake area seems to have one or more older series of sedimentary and volcanic members, and one or more younger series. Pressure from the south has resulted in the folding and thrust faulting of the younger series, and the older series may also form part of a thrust fault block.

OTHER INFORMATION ON THE EASTERLY EXTENSION OF THE GRENVILLE

Other workers have concluded that the Grenville province extends eastward as far as Lake Mistassini (24). From Lake Mistassini to the Labrador coast, a few scattered areas have been mapped by government geologists (2, 5, 7, 8, 11, 14), and some mapping has been done by geologists en-

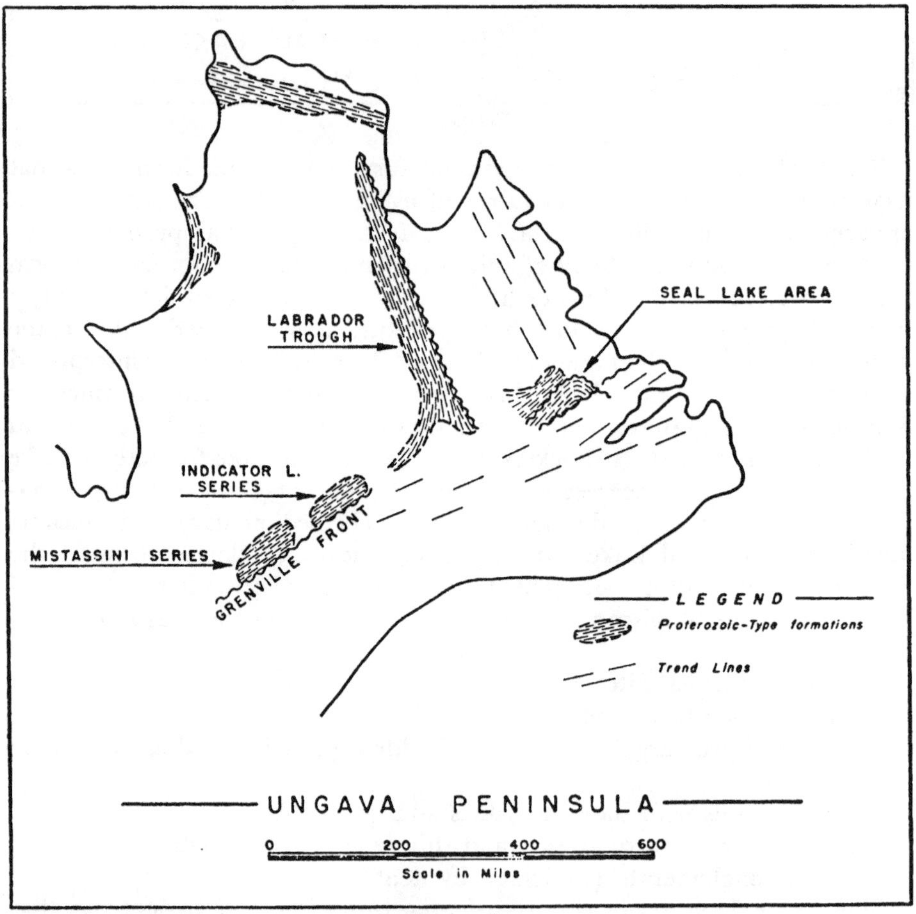

FIGURE 2

gaged in exploration work for mining companies (16, 18, 19, 20, 21, 22, 23). The available records of both government and company work have been compiled to see what light they might shed on the easterly continuation of the Grenville.

Before considering this information, some of the characteristic features of the Grenville province will be briefly enumerated. The common rock types of the province are medium to highly metamorphosed sedimentary formations that have been intruded and altered by a series of intrusive rocks ranging in composition from anorthosite to granite. Orthogneisses and paragneisses are common. The general trend of the formations is northeasterly, parallel to the Grenville front, and Gill (9) regards the province as representing the roots of a late Precambrian mountain system. The northern boundary of the province is in places marked by a zone of thrust faulting (12, 17), and remnants of younger sedimentary and volcanic formations occur at intervals adjacent to this boundary. Age determinations on minerals from the province by Wilson and others (4) indicate ages of 800 to 1100 million years.

At Lake Mistassini, the late Precambrian Mistassini series, consisting of iron formations, dolomites, and limestones, strike northeasterly and dip gently southeast. They are truncated abruptly to the southeast by garnetiferous granite gneisses, and it is probable that the contact marks a major fault between the Grenville and Superior provinces (24). The Mistassini series terminates about 20 miles north of Lake Mistassini.

The Indicator Lake series, consisting of late Precambrian sandstones, quartzites, and conglomerates, extends for a hundred miles northeast of Lake Mistassini. In the north these formations are flat-lying, but towards the south the dips become progressively steeper, and the series appears to be truncated on the southeast by faulting (16).

The Labrador trough, consisting predominantly of Proterozoic-type formations, extends in a southerly direction for a length of 600 miles. At the southern extremity, the formations become highly metamorphosed, intruded by granites, and finger out to the southwest (21, 23). This increase in metamorphism and divergence of strike, together with the presence of granite intrusives, suggests that the trough rocks at the southern end of the belt were involved in the Grenville mountain building.

From the southern end of the trough to the Seal Lake area, there is little evidence of faulting, but the general trend of the formations is northeasterly (19, 21, 22). This same trend continues past the Seal Lake area to near Cape Makkovic on the Labrador coast (19, 18, 20).

Thus between Lake Mistassini and Cape Makkovic, the general trend of the formation is northeasterly, thrust faulting occurs in places parallel to this trend, and Proterozoic-type formations occur in scattered areas. To the south the formations resemble those of the Grenville province, and so it is probable that the Mistassini–Seal Lake–Cape Makkovic line marks the northern edge of the Grenville province.

An age determination made on a radioactive mineral from north of Lake Melville gave an age of 600 million years (19).

The Area East of the Labrador Trough

North of the suggested Grenville boundary, and east of the Labrador trough, the common rock formations are metasediments and intrusives which appear to be similar to the formations of the Grenville province (19). However, the dominant trend of these Labrador formations is northerly, and this trend appears to be truncated by the northeasterly trends associated with the suggested Grenville front. Age determinations made on mineral specimens from the Labrador trough and from northeast of Seal Lake gave similar results of 1540 million years (4). The truncated trend lines, and age determinations, suggest that the formations east of the Labrador trough are not part of the Grenville province, despite lithological similarities, but belong to another geological province intermediate in age between the Grenville and Superior.

Conclusion

A great deal more field work will be needed before authoritative statements can be made about the extension of the Grenville province to the Labrador coast, but most of the evidence now available does favour this contention.

REFERENCES

(1) Beland, R. (1950). Le synclinal du lac Wakeham et la fosse du Labrador; Naturaliste canadien, vol. 77, pp. 291–304.

(2) Christie, A. M. (1951). Geology of the southern coast of Labrador from Forteau Bay to Cape Porcupine, Newfoundland; Geol. Surv., Canada, Paper 51–13.

(3) ——— (1952). Geology of the northern coast of Labrador from Grenfell Sound to Port Manvers, Newfoundland; Geol. Surv., Canada, Paper 52–22.

(4) Cumming, G. L., Wilson, J. T., Farquhar. R. M., Russell, R. D. (1955). Some dates and subdivisions of the Canadian Shield; Proc. Geol. Assoc. Canada, vol. 7, pt. 11, pp. 27–79.

(5) Eade, K. E. (1952). Preliminary map, unknown river, Labrador, Newfoundland; Geol. Surv., Canada, Paper 52–9.

(6) Engel, A. E. J. and C. G. (1953). Grenville series in the northwest Adirondack Mountains; Bull. Geol. Soc. America, vol. 64, pp. 1013–1097.

(7) Fahrig, W. (1952). Snegamook Lake map area; Geol. Surv., Canada, unpublished map.

(8) Frarey, M. J. (1952). Preliminary map, Willbob Lake, Quebec and Newfoundland; Geol. Surv., Canada, Paper 52–16.

(9) Gill, J. E. (1952). Mountain building in the Canadian Pre-Cambrian Shield; Internat. Geol. Conference, Rept. 18th Session, Great Britain, 1948, pt. XIII, pp. 97–104.

(10) Hallet, R. (1946). Geological reconnaissance of the Naskaupie Mountains; unpublished report to Dome Mines. Limited.

(11) Harrison, J. M. (1952). The Quebec Labrador Iron Belt; Geol. Surv., Canada, Paper 52–20.

(12) Johnston, W. G. Q. (1954). Geology of the Temiskaming-Grenville contact, southeast of Lake Temagami, Northern Ontario, Canada; Bull. Geol. Soc. Amer., vol. 65, pp. 1047–1074.

(13) Kranck, E. H. (1939). Bedrock geology of the seaboard region of Newfoundland Labrador; Geol. Surv., Newfoundland, Bulletin no. 19.

(14) Roscoe, S. (1952). Kasheshibaw Lake map area, Labrador; Geol. Surv., Canada, unpublished map.
(15) Scott, H. S. and Conn, H. K. (1950). General geology in the region of Harp Lake, and Canairiktok, Naskaupie, and Red Wine rivers, Labrador; unpublished report to Photographic Surveys, Limited.
(16) Tait, A. Unpublished information on Indicator Lake series.
(17) Wilson, J. T. (1949). Some major structures of the Canadian Shield; Trans. Can. Inst. Min. Met., vol. 52, pp. 231–242.
(18) American Metals Limited. Unpublished records on Labrador.
(19) British Newfoundland Corporation Limited. Unpublished records on Labrador.
(20) Frobisher Limited. Unpublished records on Labrador.
(21) Labrador Mining and Exploration Company Limited. Unpublished records on Labrador.
(22) Newfoundland and Labrador Corporation, Limited. Unpublished records on Labrador.
(23) United Dominion Mining Company. Unpublished records on Labrador and Quebec.
(24) Tectonic map of Canada, Geol. Assoc. Canada, 1950.

DISCUSSION

D. R. DERRY

Has anyone seen in any other places in the Shield gently dipping thrust planes like those just described by W. G. Robinson?

B. R. MACKAY

They are like those in the Alberta foothills.

J. M. HARRISON

W. F. Fahrig has found that granite rocks east of the Labrador trough contain granitized equivalent of trough rocks. (Similarly, the trough rocks have been traced into Grenville rocks at the southwest extremity of the trough.)

I. W. JONES

Dr. Robert Bergeron and Mr. Pierre Sauvé mapping for the Quebec Department of Mines have found that, on the east side of the Labrador trough, west of Fort Chimo, the relation between the relatively little metamorphosed Proterozoic rocks of the trough and the gneisses to the east is one of gradational change. This relation is similar to that found by other geologists of the same Department in the region east and south of Chibougamau. It would seem that in several places, then, the contact between the rocks of the so-called Grenville subprovince and those of the so-called Timiskaming subprovince is not a fault, and it is wondered what are Grenville rocks and what are not, and if there is a so-called "Grenville Front." It is urged that much caution should be exercised in discussing this problem and geologists should not generalize about it for areas which they have not seen. An attempt should be made to have a common understanding about terms that are used. It would seem, for example, that "Grenville" is being used in different senses for time, lithology, and tectonics. Indeed, as far as age is concerned, even in the papers being presented here, "Grenville" is considered as being Archean by some and Proterozoic by others—this is a wide divergence!

D. R. DERRY

Does anyone now think that the Grenville is Archean?

M. E. WILSON

I do.

J. B. MAWDSLEY

Will J. T. Wilson please tell us what is known from age determinations?

J. T. WILSON

Most of the original work in dating Grenville rocks was done by the late Dr. H. V. Ellsworth of the Geological Survey of Canada. His work was excellent and his conclusions have not been greatly changed. Recently much more work has been done at the University of Toronto by C. B. Collins, R. D. Russell, R. M. Farquhar, H. A. Shillibeer, G. L. Cumming and other graduate students. The latter two are giving us a paper shortly which will summarize this work and show that the pegmatites in the Grenville region are about one billion years old. There is considerable evidence that the Keewatin rocks were formed not less than two to two and one half billion years ago.

A. E. ENGEL

The age of a pegmatite is no indication whatsoever of the age of the sediments
it cuts. In the Mohave Desert there are Archean rocks at least 1200 million
years old overlain by Proterozoic, Paleozoic, and Tertiary sediments. These are
cut by granites indicating repeated mountain-building episodes.

I believe that a major difficulty with this subject is that practically everyone
who discusses it is either a geologist with an inadequate knowledge of physics
and chemistry or a physicist or chemist with an inadequate knowledge of
geology.

W. F. JAMES

I have crossed the Grenville front many times in a number of different places
and I believe that the rocks of the Grenville and Keewatin subprovinces were
once all similar, that they are all of the same age and were laid down at the
same time. The Grenville part was, however, made into mountains about 1000
million years ago in a manner similar to that in which the Appalachians were
formed at a still later date beyond them.

I think in these discussions we should beware of the enthusiast, who has been
defined as one who having lost sight of his goal redoubles his efforts. At the
same time it is perhaps useful to realize that geologists tend to be divisible
into two classes. On the one hand, there are the pessimists. These men, who
usually come from geological surveys and academic positions look for more
and more detail in the name of purity. They are generally very concerned with
the details of stratigraphy and they are often iconoclasts. On the other hand,
there are the optimists, who include the economic geologists as well as many
physicists and regional geologists. In conclusion I would repeat that it seems to
me that the Grenville and Keewatin rocks are the same and that we should
use the term "groups" in referring to suites of Precambrian rocks.

THE GRENVILLE REGION OF ONTARIO

D. F. Hewitt

THE GRENVILLE SUBPROVINCE occupies southeastern Ontario between the Grenville "front" on the north and the Paleozoic contact on the south. The Grenville "front" extends from Killarney at the north end of Georgian Bay, to the south end of Lake Timiskaming. The southern boundary of the subprovince, the Paleozoic contact, extends from the south end of Georgian Bay to Kingston on Lake Ontario.

The Grenville series of Precambrian metasediments was first recognized and named by Sir Wm. Logan near the town of Grenville in Argenteuil County, Quebec, on the Ottawa River, and this is the type locality. Early reports of the Geological Survey of Canada by Logan, Vennor, MacFarlane, and Selwyn, prior to 1900, indicated that metasediments of the Grenville type extended westward from the type area into southeastern Ontario, and subsequently nearly all the Precambrian metasediments of southeastern Ontario were correlated with the Grenville.

During the period before 1900, work by MacFarlane and Vennor in the Madoc area of eastern Ontario indicated that there was a series of Precambrian sedimentary rocks present in that area, consisting of conglomerate, argillite, schist, and well-bedded blue limestones, which were much less metamorphosed than the Grenville series of the type area. These sediments were called the Hastings series, from their type locality in the southern part of Hastings County, Ontario. However, apart from their apparent lower grade of metamorphism, and the presence of conglomerate in the section, no good criteria as to age or stratigraphic position with respect to the Grenville were given, when the series was recognized by the early workers. Thus the controversy as to whether the Hastings series was a less metamorphosed facies of the Grenville or a younger series resting unconformably on the Grenville was born, and is still not satisfactorily solved.

At present the general field usage in eastern Ontario is to refer the paragneisses, amphibolites, quartzites, and schists of higher metamorphic grade to the Grenville series, or to call them metasediments of the "Grenville-type." Conglomerates and the younger less metamorphosed schists, argillites, and sandstones, in which original sedimentary features such as crossbedding are well preserved, are generally referred to as "Hastings-type" sediments. This usage is based on lithology and grade of metamorphism, the same criteria the early workers used in originally distinguishing Hastings from Grenville.

Work in eastern Ontario by M. E. Wilson (1933, 1940), and more recently by B. L. Smith (1951) and C. A. Burns (1951), has indicated that grade of metamorphism and lithology are not valid criteria for distinguishing Hastings from Grenville. Important conglomerate members are found in the Hastings-Grenville sequence at three different stratigraphic horizons at least. There appears to be proof of the presence of a series of sediments and volcanics which are younger than the bulk of the Grenville series. Granitized sediments of high metamorphic grade which would be described as "Grenville-type," under lithologic field usage, have been found to be stratigraphically above "Hastings-type" conglomerates and blue, well-bedded, limestones. It appears that there will have to be a re-definition of the terms "Grenville" and "Hastings" in eastern Ontario, on a structural and stratigraphic basis.

The age of the Grenville has long been in dispute. Early workers correlated the Grenville with the Huronian. Miller and Knight (1914) correlated the Hastings with the Timiskaming and placed the Grenville as post-Keewatin, pre-Timiskaming. Recent radioactive age work on intrusives which cut the Grenville has indicated that the oldest intrusives examined, the nepheline syenites, have an age of 1370 ± 120 million years.[1] Pegmatites and granites in the Wilberforce area will give ages in the range from 1020 to 1100 million years, while some late pegmatites at Hybla and Britt, Ontario, give ages as young as 780 million years. These ages on intrusive and plutonic rocks which are younger than the Grenville place an upper limit of about 1400 million years for the age of the Grenville.

The following are the main features of the geologic history in the Grenville region of eastern Ontario as they appear to the writer:

1. Sediments of the Grenville type, not volcanics, are the earliest rocks present in eastern Ontario.

2. Volcanics occur within the Grenville series, towards the upper part of the series. There are no pre-Grenville volcanics.

3. There is no evidence of a basement for the Grenville. The so-called Laurentian granites and granite gneisses invade and replace the Grenville series, and also invade and replace "Hastings-type" sediments.

4. Younger series of sediments and volcanics, including some "Hastings-type" conglomerates, lie above or at the top of the Grenville sequence, but no major stratigraphic break has yet been proven. Conglomerate occurs at more than one stratigraphic horizon.

5. A major orogeny followed the period of sedimentation and volcanism. Intrusion, replacement, metasomatism, and metamorphism had profound effects, varying in intensity from area to area within eastern Ontario. Some general features are common to the Grenville of the whole region, but central eastern Ontario can be divided into structural units with some distinctive features. In parts of the Grenville region we are dealing essentially with a high grade metamorphic terrane which has suffered deep burial.

[1]H. A. Shillibeer, personal communication.

Other parts show a relatively low grade metamorphism with preservation of original structures and textures in sedimentary, volcanic, and intrusive rocks.

Progress of Geological Mapping in the Grenville Region of Ontario

The accompanying map, Figure 1, indicates the areas in which geological mapping has been completed in eastern Ontario. Mapping on the scale of one mile to the inch is incomplete, and many geological problems are not yet solved. Progress in interpretation of the geological history of the area has depended on the progress of the geological mapping, and there has been much controversy on various problems of the Grenville.

Pre-1900

In the early period of geological exploration prior to 1900 work was carried out by the Geological Survey of Canada and the annual reports by Logan, Vennor, and Selwyn covering this work are well known. The early concept of geological relations in eastern Ontario visualized a great metamorphic series of Precambrian age known as the Laurentian, divided into a lower division, the Ottawa gneisses, and an upper division, the Grenville series of Logan. The Ottawa gneisses consisted of a complex of hybrid granite and granitized gneisses, which was regarded as the "fundamental" gneiss basement. Following Lyell, most of the early workers interpreted the foliated granitic gneisses as being of sedimentary origin, regarding foliation as indisputable evidence of this fact. This bias is still held by many geologists today.

The Grenville series of Logan, characterized by crystalline limestone, schist, and paragneisses, was believed to rest on the fundamental gneiss basement. The Hastings series of Hastings County, Ontario, was interpreted by Vennor and Selwyn[2] as conformable with the Grenville, and of Huronian age. There was some controversy as to whether the Laurentian Ottawa gneiss was the basement on which the Grenville rested unconformably, or whether the Ottawa gneiss intruded the Grenville. It was suggested that the Ottawa gneiss of the basement may have been re-fused and intruded into the Grenville.

Adams and Barlow (1910)

F. D. Adams and A. E. Barlow in 1910 published their memoir on the geology of the Haliburton-Bancroft area, embodying results of their field work in eastern Ontario from about 1892 to 1902. Adams concluded that the Ottawa "fundamental" gneiss of the lower Laurentian invaded the Hastings-Grenville sediments. The Grenville was believed to grade into a less altered phase, the Hastings series. In the Madoc area these geologists found that the Hastings consisted of two unconformable series and concluded that therefore so does the Grenville.

[2]Geological Survey of Canada, Reports of Progress, 1863–66, 1877–78.

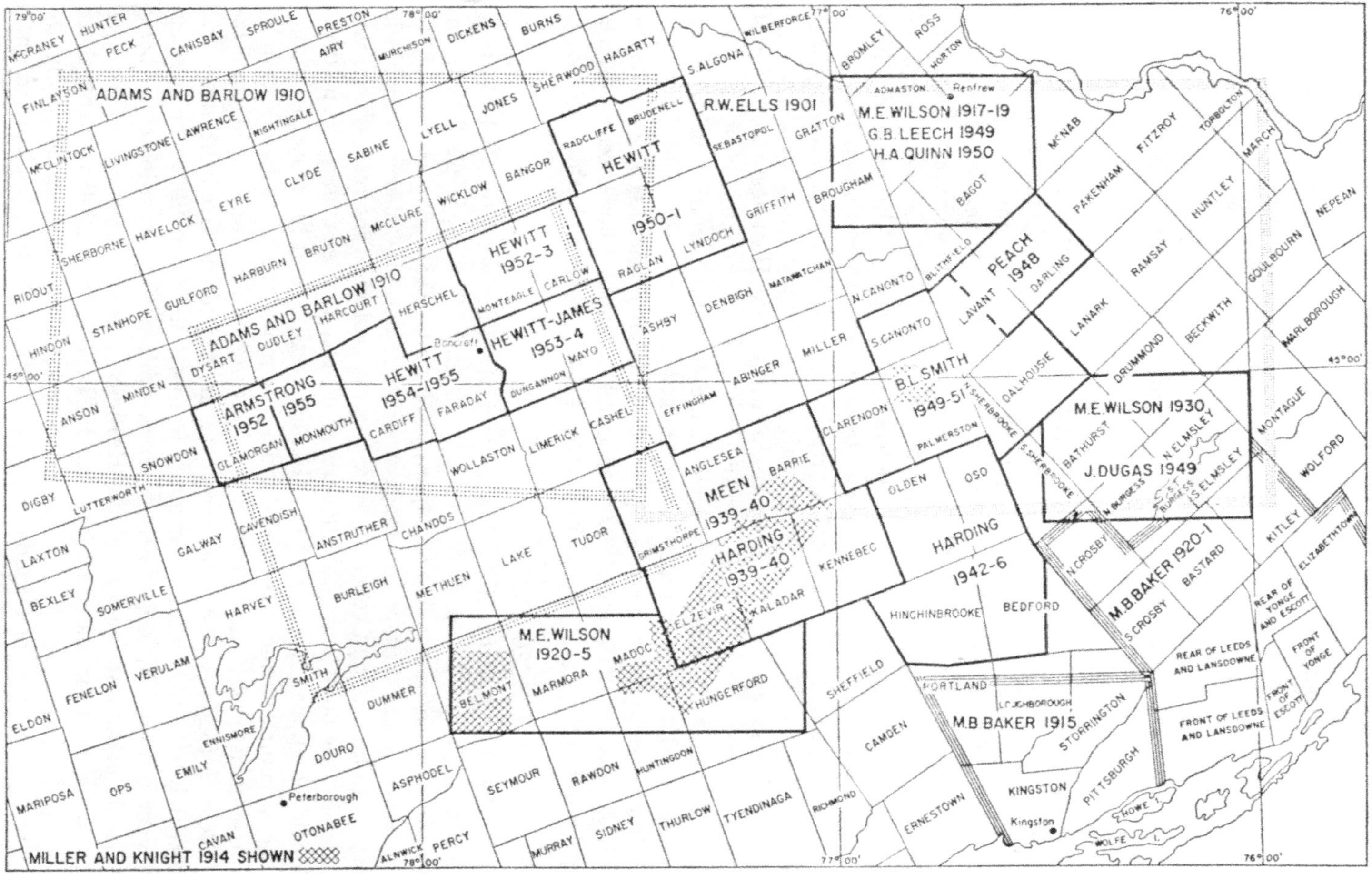

Figure 1. Geological mapping in Eastern Ontario.

The Haliburton sheet, map 708, on the scale of 4 miles to the inch, and the Bancroft sheet, map 770, on the scale of 2 miles to the inch, which accompanied Adams's and Barlow's report, have been very useful to geologists working in eastern Ontario, and in several townships are still the only geological maps available.

R. W. Ells completed the Perth sheet, map 119, and the Pembroke sheet, map 660, which were published in 1901 and 1906 respectively.

Miller and Knight (1914)

W. G. Miller and C. W. Knight of the Ontario Bureau of Mines in 1914 published the results of their mapping in a number of selected areas in eastern Ontario. They concluded that the oldest rocks present were Keewatin volcanics. The Keewatin was overlain conformably by the Grenville series; this was in turn overlain unconformably by the Hastings series which they correlated with the Timiskaming. The Keewatin and Grenville were invaded by Laurentian granite. The Hastings series, at the base of which was a conglomerate containing granite pebbles and limestone pebbles containing "eozoon," overlies the Laurentian granite with unconformity, but no place where the two are in unconformable contact was found.

The Moira granite was believed to intrude the Hastings and to be of Algoman age. However, no intrusive relations between the Hastings sediments and the Moira granite were found, the interpretation being based largely on difference in grade of metamorphism.

Subsequently map sheets of parts of Frontenac and Leeds counties by M. B. Baker (1916 and 1922 respectively), and a map sheet of Brockville-Mallorytown area by J. F. Wright (1923), were published.

M. E. Wilson (1920–1925)

In 1940 map sheets of the Madoc and Marmora areas by M. E. Wilson, based on field work done between 1920 and 1925, were published, unfortunately without any accompanying report. These map sheets can be regarded as the first of the modern series of one mile sheets available in eastern Ontario, and cover an area of great geological interest. Wilson was the first to point out that modern structural and stratigraphic methods could be applied in parts of the Grenville region, and his maps delineated the structural elements of the Madoc-Marmora area; the best known of these structures is the Clare River syncline. Wilson concluded that the volcanics found in the area were near the top of the Grenville series and occur interbedded with sediments. He found the Grenville to be overlain unconformably by the Hastings series, with a basal conglomerate being usually present at the base of the Hastings.

Work begun in the Perth area by M. E. Wilson in 1930 was completed by J. Dugas in 1949, and the Perth map sheet was published by the Geological Survey of Canada in 1950 as Paper 50–29. Work carried on in 1917 to 1919 by M. E. Wilson in the Renfrew map area was completed in

1949 and 1950 by G. B. Leech and H. A. Quinn, and this map sheet was published by the Geological Survey of Canada in 1952 as Paper 51–27.

Ontario Department of Mines (1939–1954)

In 1939 and 1940 V. B. Meen mapped Grimsthorpe, Elzevir, Anglesea, and Barrie townships, and W. D. Harding mapped Kaladar and Kennebec townships. The results of this field work were published as Map 51d, Grimsthorpe-Kennebec area, on the scale of one mile to the inch. Both Meen and Harding found difficulty in separating the Grenville from the Hastings in their respective townships. The well-known Flinton conglomerate band is shown as Hastings, but the Kaladar conglomerate and the eastward extension of the younger beds in the Clare River syncline are shown as Grenville on the Grimsthorpe-Kennebec sheet. However, the criteria used for separating Hastings from Grenville are not clear.

Between 1942 and 1946 W. D. Harding completed the mapping of Olden, Oso, Hinchinbrooke, and part of Bedford townships in Frontenac County, and map 1947–5, "Olden-Bedford Area," on the scale of one mile to the inch, was published with a report covering this area. The grade of metamorphism in the Olden-Bedford area is higher than that in the Grimsthorpe-Kennebec area, and increases to the south and east from the Clare River syncline. All the sediments mapped in this area were referred to as the Grenville series by Harding.

In 1941, 1942, and 1943 J. Satterly examined mineral occurrences in eastern Ontario and accompanying his reports, compilation maps of the general geology in parts of the Districts of Parry Sound and Muskoka, and of Haliburton and Renfrew counties were published. A similar report on mineral occurrences and a geological compilation map of North Hastings County were completed by J. E. Thomson in 1942.

In 1948 P. A. Peach mapped Darling Township and a portion of Lavant Township, Lanark County. In 1949, 1950, and 1951 B. L. Smith continued this mapping, on the scale of one mile to the inch, and completed Lavant, Dalhousie, North Sherbrooke, Palmerston, and Clarendon townships to the south and west of the Lavant-Darling area. Preliminary reports on a part of this work have been published. Smith (1951) recognizes a major northeast trending fault, the Fernleigh-Clyde fault, which extends from Mazinaw Lake eastward across Barrie, Clarendon, Palmerston, and Lavant townships to the White Lake granite gneiss in southern Bagot Township. This fault separates the more highly metamorphosed rocks to the north from the less metamorphosed rocks of the Kaladar-Dalhousie trough to the south.

South of the fault Smith recognizes an older series consisting of metavolcanics, crystalline limestone and dolomite, paragneiss, metagreywacke, quartzite and biotite and hornblende schist. The older series is overlain by Hastings-type conglomerate in the Ompah syncline in Palmerston Township. Above the conglomerate are quartzite, argillite, schists, and some limestone. This succession is overlain by younger biotite and hornblende schists which are interpreted as metavolcanic in part. These metavolcanics are overlain by less metamorphosed blue limestones.

In 1950 and 1951 D. F. Hewitt completed a one mile to the inch sheet of Radcliffe, Raglan, Brudenell, and Lyndoch townships, Renfrew County. In 1952 and 1953 Monteagle and Carlow townships in Hastings County were also completed, and will be published, together with succeeding maps in this series, on the scale of one-half mile to the inch. In 1953 and 1954 D. F. Hewitt and W. James completed Dungannon and Mayo townships, Hastings County. In 1951 H. S. Armstrong mapped Glamorgan Township, Haliburton County. The mapping of Monmouth and Cardiff townships in Haliburton, and Faraday Township in Hastings County, is planned to complete this series of map sheets.

Other isolated areas in the Grenville region of Ontario have also been mapped. In 1943 W. D. Harding mapped parts of Mattawan and Olrig townships, District of Nipissing, north of the Mattawa River. In 1953 J. Satterly mapped Lount Township, District of Parry Sound.

Remapping of part of the Kaladar-Kennebec area was carried out by C. A. Burns, under the auspices of the Geological Survey of Canada in 1950.

General Geology and Structure

Table I is a tentative geological timetable for the Grenville region of central eastern Ontario.

Elements of Regional Structure

1. *Haliburton, Hastings, and Madawaska Highlands*

The Haliburton, Hastings, and Madawaska Highlands, lying to the north and west (see Figure 2), consists of terranes of high grade metamorphic gneisses, characterized by rocks of the amphibolite and granulite metamorphic facies. For the most part intrusives are concordant and there is much evidence of granitization and metasomatism, with the formation of hybrid rocks by these processes. Destruction of original sedimentary and volcanic features in the rocks prevents application of normal stratigraphic methods. Sediments, for the most part, occur as relicts and pendants in areas of hybrid granite gneiss.

In the northwestern part of the area the Dysart and Glamorgan bodies of granite gneiss are separated from the main mass of hybrid granite gneiss by fingers of metasediments, mainly crystalline limestone. To the south of the Glamorgan granite gneiss lies the Cavendish-Monmouth sedimentary trough, consisting mainly of crystalline limestone, paragneiss, and quartzite. Within this sedimentary trough are the Gooderham and Tory Hill alkaline syenite gneisses and the Green Mountain gabbro-metagabbro complex.

To the southeast there are a series of four granite and hybrid granite gneiss bodies, termed "batholiths" by Adams and Barlow. A study of the

aerial photographs covering the most southerly of these bodies, the Burleigh granite gneiss, indicates that it consists of gneissic rocks which are
folded into a series of three anticlines and synclines pitching south. The
easternmost syncline within the Burleigh body lies just west of No. 28 highway and is overturned to the northwest. To the south the Burleigh body is
overlain by Paleozoic limestones.

The Anstruther granite gneiss, which lies to the north, consists of a large
oval body of granite and hybrid granite gneiss almost surrounded by meta-

TENTATIVE GEOLOGICAL TIMETABLE

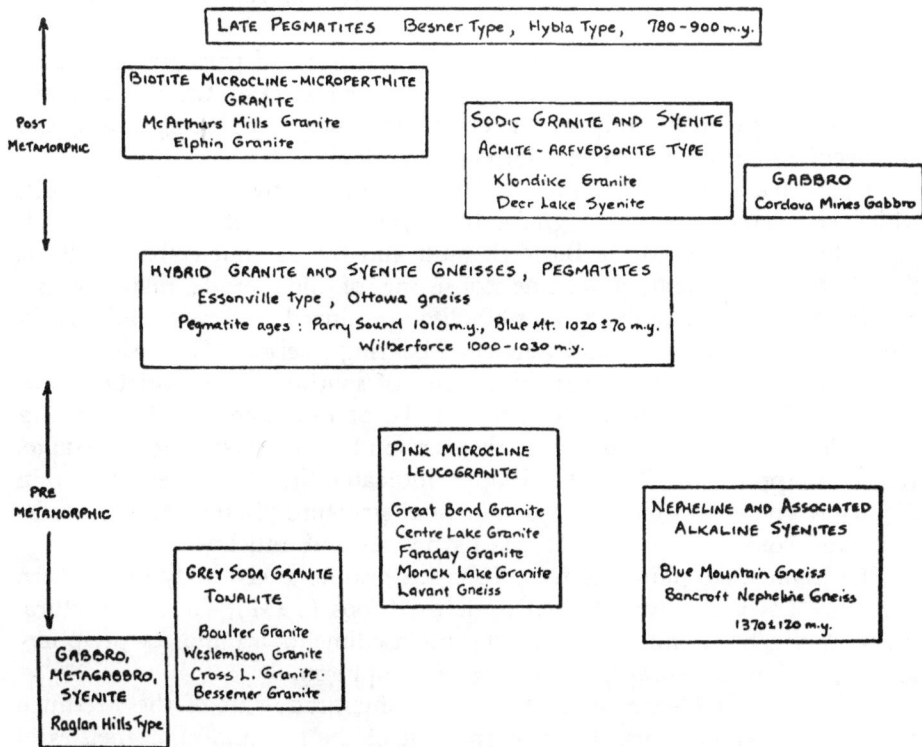

LATE PEGMATITES Besner Type , Hybla Type ,   780 - 900 m.y.

POST
METAMORPHIC

BIOTITE MICROCLINE - MICROPERTHITE
GRANITE
McArthurs Mills Granite
Elphin Granite

SODIC GRANITE AND SYENITE
ACMITE - ARFVEDSONITE TYPE
Klondike Granite
Deer Lake Syenite

GABBRO
Cordova Mines Gabbro

HYBRID GRANITE AND SYENITE GNEISSES , PEGMATITES
Essonville type , Ottawa gneiss
Pegmatite ages : Parry Sound 1010 m.y., Blue Mt. 1020 ± 70 m.y.
Wilberforce 1000 - 1030 m.y.

PRE
METAMORPHIC

PINK MICROCLINE
LEUCOGRANITE
Great Bend Granite
Centre Lake Granite
Faraday Granite
Monck Lake Granite
Lavant Gneiss

NEPHELINE AND ASSOCIATED
ALKALINE SYENITES
Blue Mountain Gneiss
Bancroft Nepheline Gneiss
1370 ± 120 m.y.

GREY SODA GRANITE
TONALITE
Boulter Granite
Weslemkoon Granite
Cross L. Granite.
Bessemer Granite

GABBRO,
METAGABBRO,
SYENITE
Raglan Hills Type

INTRUSIVE   CONTACTS

GRENVILLE SERIES	HASTINGS GROUP	5	Blue limestones, quartzite
		4	Volcanics, hornblende schist
		3	Conglomerate (pebbles of qtz.,qtzite, ls.,argillite,granite) Argillite, quartzite, schist
	GRENVILLE GROUP	2	Volcanics, pyroclastics, conglomerate, limestone
		1	Limestone, paragneiss, amphibolite, quartzite

TABLE I

sediments. The foliation within the body shows a double domal structure. Narrow relict bands of limestone and amphibolite occur well within the hybrid granite gneiss, parallel to the contacts.

To the north, and separated from the Anstruther gneiss by bands of limestone and paragneiss, is the Cheddar granite. This body is almost circular in shape and consists of fairly massive pink biotite granite with a border which is pegmatitic in places. The Cheddar granite consists of two domes, a large one to the north and a small one to the south, with a re-entrant of sediments extending westward into the east side of the body. The structure is not domal, as the sediments on the north side of the Cheddar granite dip steeply to the south towards the centre of the granite body.

The Cardiff body to the north consists of a dome of paragneiss, amphibolite, hybrid granite, and syenite gneiss, centred at Deer Lake. The main features of the Cardiff body are the concordant sheets of leucogranite gneiss and syenite gneiss lying on the flanks of the dome.

The Burleigh granite gneiss is cut off on the east by the Burleigh fault, which separates the granite gneiss from metasedimentary gneisses which lie to the east. Between the Burleigh fault and the Methuen "batholith" is the Methuen segment, a terrane consisting of high grade metamorphic gneisses including paragneiss, amphibolite, granitized and syenitized hybrid gneisses, syenite, granite, and nepheline-bearing gneisses. The structure in this segment appears to consist of a series of south-pitching anticlines and synclines. On the north side of Stony Lake at the eastern end there is a small dome of granite gneiss nearly surrounded by crystalline limestone. Recent mapping by C. V. G. Phipps indicates that the Blue Mountain nepheline gneiss occupies a major synclinal structure pitching to the southwest, and consists of a series of small synclines and anticlines.

The Loon Lake pluton, a funnel-shaped granite gneiss body in southern Chandos Township, was studied by Ernst Cloos (1934). Cloos' structural study of the body indicated that the metasediments around its periphery all dipped inward towards the centre of the pluton.

There is a notable decrease of metamorphic grade east of the Methuen "batholith," which marks the eastern limit of the metamorphic gneisses of the highland area and the western limits of the Hastings basin.

2. Hastings Basin

The Hastings basin occupies the southern part of Hastings County and is terminated on the west by the Methuen "batholith," on the north by the McArthurs Mills line which may be a major fault zone, and on the east by the Weslemkoon and Elzevir batholiths.

The Hastings basin consists of a terrane of low to intermediate grade of metamorphism, including schists, argillites, well-bedded blue limestones, crystalline limestones, and volcanics. High grade metamorphic gneisses and large areas of hybrid granitized gneisses are, for the most part, lacking.

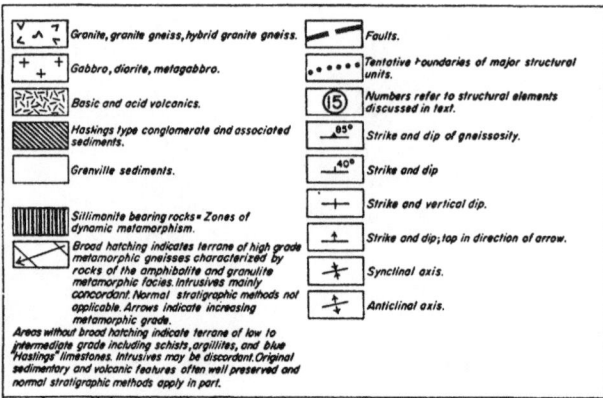

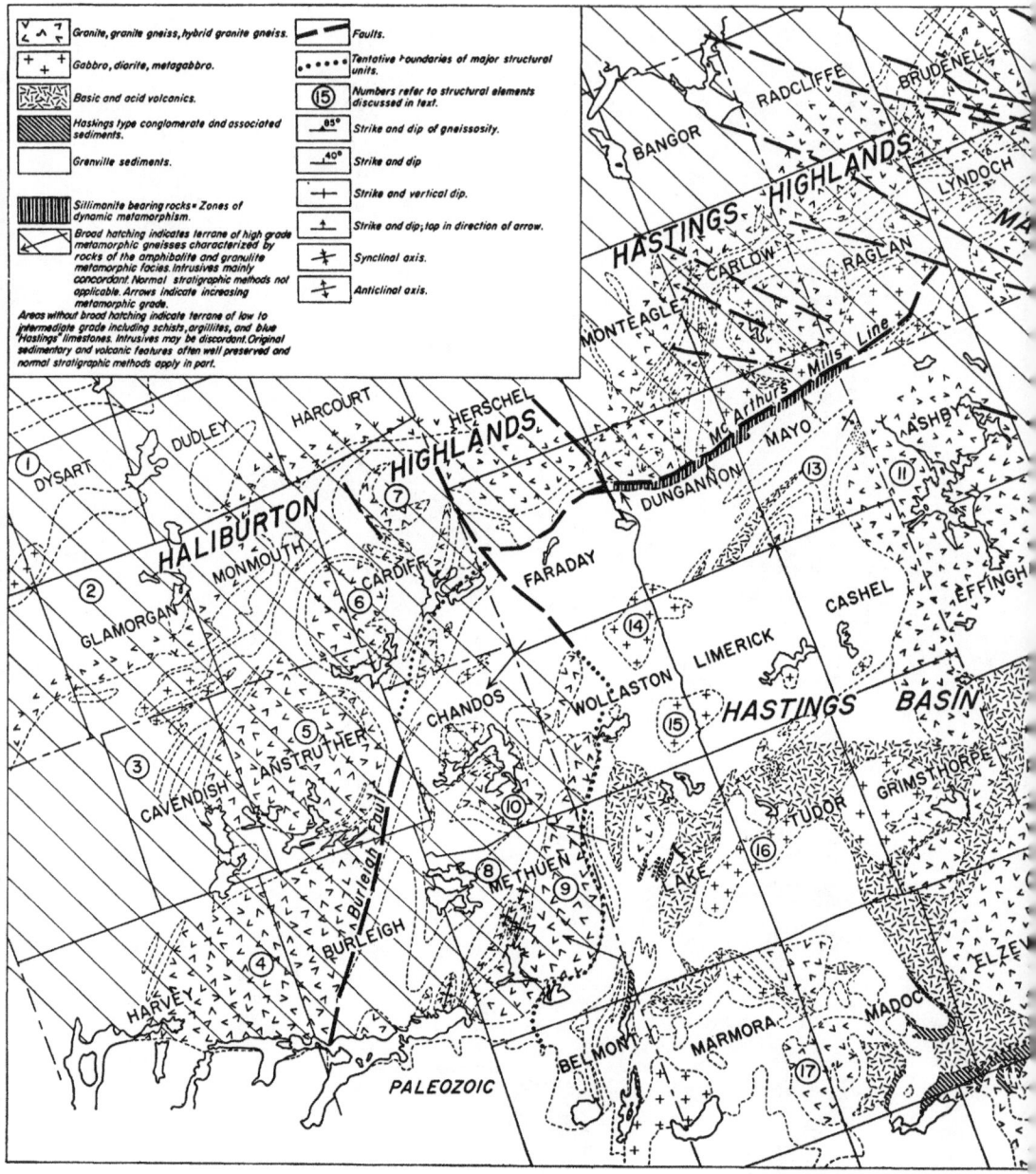

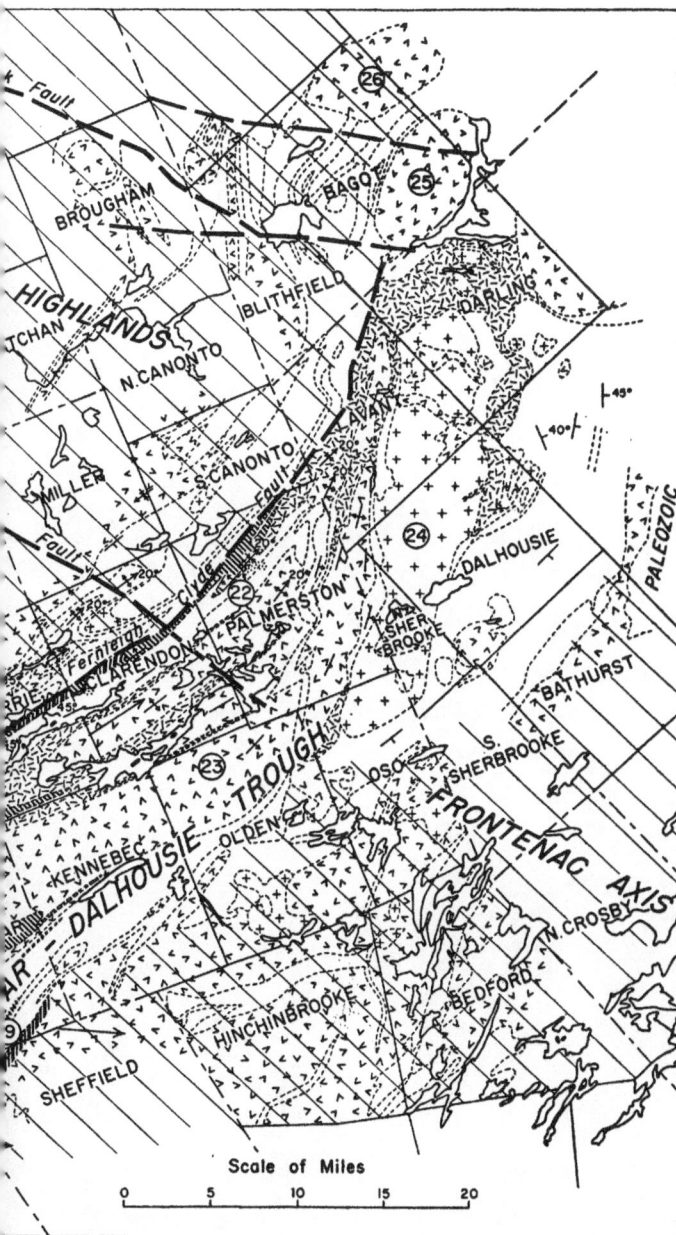

FIGURE 2. The Grenville subprovince, eastern Ontario.

ELEMENTS OF REGIONAL STRUCTURE

1. Dysart granite gneiss
2. Glamorgan granite gneiss
3. Cavendish-Monmouth trough
4. Burleigh granite gneiss
5. Anstruther granite gneiss
6. Cheddar granite
7. Cardiff dome
8. Methuen segment
9. Methuen batholith
10. Loon Lake pluton
11. Weslemkoon batholith
12. Elzevir batholith
13. Mayo anticline
14. Umfraville gabbro
15. Thanet gabbro
16. Tudor gabbro
17. Deloro granite
18. Clare River syncline
19. Clare River conglomerate
20. Kaladar conglomerate
21. Flinton conglomerate
22. Ompah syncline, Ompah conglomerate
23. Cross Lake gneiss, Cross Lake anticline
24. Dalhousie gabbro complex
25. White Lake granite gneiss
26. Hond Lake granite gneiss

Sedimentary and volcanic structures such as crossbedding, grain gradation, and pillows, are frequently well preserved, and normal structural and stratigraphic methods may be applied to solve the geologic structures within most of the area. This contrasts markedly with conditions in the Hastings highlands north of the McArthur Mills line where the rocks are largely metamorphic gneisses in which top determinations cannot be made.

In southern Mayo and Dungannon townships the structure is a major anticline pitching to the northeast. The sedimentary sequence consists of a series (from bottom to top) of crystalline limestones and impure argillaceous limestones (often well-bedded), rusty schists, rhyolite, limestone and feather amphibolite, basic volcanics, limestone and feather amphibolite. This series, which was tentatively called the Mayo group by the writer, has a thickness of at least 12,000 feet, and appears to be cut off to the north at the McArthurs Mills line which may be a major east-west fault zone. These sediments are intruded by gabbro bodies such as the Umfraville gabbro, and by the Bessemer, McArthurs Mills, and Weslemkoon granites.

Wollaston, Limerick, Cashel, Tudor, and Lake townships have not been remapped since Adams' and Barlow's time and the structure is not well known. Limestones occupy a large portion of Wollaston, Limerick, and Cashel townships. Basic volcanics, including good pillow lavas, occupy parts of Tudor and Grimsthorpe townships. The Grimsthorpe, Thanet, and Tudor gabbro bodies intrude volcanics and sediments, as do several granite and syenite bodies.

Belmont, Marmora, and Madoc townships are included on M. E. Wilson's Marmora and Madoc map sheets. Although more detailed structural work is necessary in the Belmont Lake area, it appears likely that the volcanics at Belmont Lake are younger than the Oak Lake schists which lie to the west, and that the Hastings-type conglomerate at Belmont Lake is not an infolded remnant, but part of an east-facing sedimentary sequence. Other interesting features of this area are the Madoc andesite, and the acid volcanics and pyroclastics. The main acid intrusives are the Delora and Moira granites.

The structure within the Hastings basin is heterogeneous and does not conform to the northeast trend of folding found in the Kaladar-Dalhousie trough to the east.

3. Kaladar-Dalhousie Trough

The Kaladar-Dalhousie trough comprises a series of northeast trending folds limited on the north by the Elzevir batholith and the Fernleigh-Clyde fault, which extends northeast to White Lake, and on the south by the higher grade metamorphic gneisses of Sheffield, Hinchinbrooke, southern Olden and Oso, south Sherbrooke, and Bathurst townships.

Within this trough the sediments are somewhat lower in metamorphic grade than to the north and south, but are intruded, injected, and replaced by granitic gneisses. Sedimentary features such as crossbedding and grain gradation are preserved in many places in the sediments, and pillow tops may be found in the volcanics. The Flinton, Kaladar, Clare River, and Ompah conglomerates of the "Hastings-type" occur in the Kaladar-Dalhousie trough.

One of the main structural features is the Clare River syncline which trends and pitches northeast through Hungerford, Sheffield, Kaladar, and Kennebec townships. M. E. Wilson indicates a conglomerate horizon in the middle of the Clare River section, which consists of volcanics, argillite, and limestone, and refers the beds above the conglomerate to the Hastings, and the beds below the conglomerate to the Grenville. The grade of metamorphism increases to the northeast and granitic gneisses appear within the sequence.

The Kaladar conglomerate lies on the north limb of the syncline and separated from the predominantly metasedimentary sequence of the central part of the' Clare River syncline by about a mile of granitized gneisses. C. A. Burns (1951, p. 8) suggests that the Kaladar conglomerate lies stratigraphically below the other metasediments of the Clare River syncline and states that "rocks previously mapped as Grenville series are stratigraphically above conglomerate. Furthermore conglomerate occurs at two and possibly three horizons. These conglomerates cannot be at the base of the Hastings series unless a single bed is repeated by faulting, and no evidence was found to suggest such faulting."

Between the Kaladar conglomerate and the Flinton conglomerate which lies to the northwest, there is another belt of grey and pink granitic gneiss. The Flinton and Kaladar conglomerates merge in southeastern Elzevir Township and the structure may be anticlinal. B. L. Smith (1951), in studying the Cross Lake grey granitic gneiss which forms the eastward extension of the granite gneiss which lies between the two conglomerates, interprets the structure as anticlinal. The pink and grey granitic gneisses which occur interbanded with the metasediments are in part of replacement origin. Undoubted sedimentary conglomerate bands in the Kaladar area can be traced along strike into granitized gneiss with relict pebbles and finally into granitic gneiss.

Referring to the Clare River syncline, Burns (1951, p. 9) states:

Without satisfactory criteria to separate these metamorphic rocks into previously used divisions, the writer is adopting new nomenclature. The rocks are divided into the Tweed group and an underlying Kaladar group on the basis of structure and stratigraphy. A conglomerate member marks the base of each group. A third belt of rocks, the Flinton group, contains conglomerate, but pending complete structure of the Northbrook (Cross Lake) gneiss, is not definitely correlated with either the Kaladar or Tweed groups.

The Cross Lake anticline occupies south-central Clarendon and Palmerston townships, and plunges northeast. To the north, just south of the Fernleigh-Clyde fault, are the younger sedimentary series described by B. L. Smith. The conglomerate member which forms the base of this series is best seen in the Ompah syncline in Palmerston Township. Excellent crossbedding occurs in the quartzite which lies above the conglomerate in northern Palmerston Township.

One of the main features of the eastern end of the trough is the large diorite-gabbro complex, 24 miles long, extending in a northerly direction from northern Oso Township to western Darling Township. The limestones to the north and south of this body are well-bedded blue limestones of the "Hastings-type." Similarly, farther to the north on the south side of White Lake there are limestones and schists of much lower metamorphic grade than the metasedimentary gneisses which lie north of the White Lake granite gneiss. It appears that the younger sediments at the northeastern end of the Kaladar-Dalhousie trough are lower in metamorphic grade than those to the north and south. H. A. Quinn (1951) has suggested that this belt of younger sediments may extend northeast to join the Bristol series in McNab Township.

4. Frontenac Axis

To the south of the Kaladar-Dalhousie trough is another terrane of high grade metamorphic gneisses which extends to the south forming the Frontenac axis, a bridge between the Precambrian areas of eastern Ontario and the Adirondacks.

SOME FEATURES AND PROBLEMS OF THE GRENVILLE

1. Metamorphism

The main metamorphic processes which have affected the rocks of the Grenville region are *regional metamorphism* and *metasomatism*. *Contact metamorphism* associated with intrusives, and *dynamic metamorphism* associated with zones of shearing and faulting, have been effective locally.

Within the areas of the Haliburton-Hastings-Madawaska highlands studied by the writer the metasedimentary rocks consist predominantly of metamorphic gneisses of the amphibolite and pyroxene granulite metamorphic facies. These gneisses have been produced by widespread, high grade, regional metamorphism under conditions of high pressure and temperature and deformation. The writer believes that these metamorphic gneisses are the product of deep zone regional metamorphism. The large percentage of granitic, syenitic, and hybrid gneisses present in the Highlands indicates that various processes of *metasomatism*, including granitization, syenitization, and nephelinization, as well as the formation of metasomatic amphibolite and pyroxenite, have been active.

In considering the Haliburton-Hastings-Madawaska Highlands the problem arises as to whether or not the high grade of metamorphism is due to regional intrusion of much granitic magma. Adams and Barlow (1897) point out that the grade of metamorphism increases from the "Hastings" of the Madoc area to the "Grenville" of the Bancroft area and interpret this increase as due to the large percentage of granitic intrusives in the northern area. They state (1897, p. 177), that "the investigations so far indicate that in the region in question the Hastings series would seem to represent the Grenville series in a less altered form. In other words, the Hastings series, when invaded, disintegrated, fretted away and intensely metamorphosed by and mixed up with the underlying magma of the Fundamental Gneiss, constitutes what has elsewhere been termed the Grenville Series."

The writer suggests that the extensive granitic gneisses in the highland area are not the direct cause of high grade regional metamorphism. Metasomatic processes such as granitization are active in the zone of fusion where deep zone regional metamorphism takes place. Thus the association of hybrid granite gneiss and high grade metamorphic gneiss is compatible. As Walton (1953, 1955) has suggested in the eastern Adirondacks, "the level of regional metamorphism attained was such that the stable mineral assemblages in the metamorphic rocks were in equilibrium with P-T conditions at which granitic fluids were also stable, and that such fluids arose and produced both magmas and metasomatism wherever the appropriate components [conditions] were present." Hybrid granite gneiss and high grade metamorphic gneiss are not necessarily *cause* and *effect*, but rather they are both results of P-T conditions in deep zone regional metamorphism. They are facies in harmony with each other.

It has not been possible in the highland area so far mapped to map zones of progressive regional metamorphism, perhaps because of the uniform high grade. It does appear possible to discern zones of progressive metamorphism in the Hastings basin and Kaladar-Dalhousie trough.

As indicated on the tentative geologic timetable given earlier, some of the intrusive and hybrid plutonic rocks are pre-metamorphic. Indications of high grade metamorphism in these rocks include the development of granulitic fabric in granite gneiss, with a pronounced lineation of quartz lentils on the plane of foliation; granoblastic recrystallization of the mineral constituents with destruction of any original eutectic perthites; the alteration of gabbro to metagabbro (which is essentially a granoblastic hornblende-plagioclase gneiss often with clotted aggregates of recrystallized hornblende), etc.

A. METAMORPHIC ASSEMBLAGES

Table II gives in outline some of the main metamorphic assemblages resulting from regional metamorphism and metasomatism of Grenville sediments. Within the highland area, paragneiss, amphibolite, quartzite, and marble are the main rock types.

(i) The following assemblages are developed under regional metamorphism:

Paragneiss. The typical paragneiss consists of the assemblage *biotite-quartz-plagioclase* with or without muscovite, microcline, sillimanite, garnet, and kyanite. It is probably derived through regional metamorphism of a sandy shale, although as Engel and Engel (1953) point out, the problem of the high soda to potash ratio is a puzzling one.

METAMORPHIC ASSEMBLAGES OF THE GRENVILLE

PROCESS	INITIAL COMP^N	LOW-INTERMED. GRADE (Argillite-slate-phyllite-schist)	HIGH GRADE (Granulite - amphibolite - metamorphic gneisses)
REGIONAL METAMORPHISM (ARGILLACEOUS)	Sandy shale	Sandy argillite	PARAGNEISS = Biotite - quartz - plagioclase gneiss ± muscovite, microcline ± sillimanite, kyanite, garnet
	Shale	Argillite → Phyllite → Slate	Hornblende - biotite - quartz - plagioclase gneiss
	Limy shale	Limy shale	Hornblende - biotite - plagioclase gneiss (BIOTITE AMPHIBOLITE)
(CALCAREOUS)	Shaly limestone	Shaly marble	AMPHIBOLITE: Hornblende - plagioclase gneiss ± microcline ± garnet ± scapolite
	Limestone	Marble	Clinopyroxene (diopside) - hornblende - plagioclase gneiss ± biotite ± orthopyroxene (hypersthene) ± garnet ± scapolite
	Sandy limestone	Sandy marble	MARBLE: Marble with silicates, particularly mica : calcite (dolomite) - phlogopite - diopside - tremolite
	Limy sandstone		Marble with silicates, particularly diopside, tremolite : calcite - diopside - tremolite; quartz - diopside - tremolite
(SANDY)	Sandstone	Quartzite	Quartz - biotite - plagioclase ± microcline (QUARTZO - FELDSPATHIC GNEISS)
	Shaly sandstone		

METASOMATIC PROCESS	ORIGINAL COMPOSITION	INTRODUCED MATERIAL	FEATURES : Changes in mineralogy and texture
Granitization and Syenitization	PARAGNEISS	K_2O+	Biotite - quartz - plagioclase → MIGMATITE → FELDSPAR PORPHYROBLASTS Wrapping around or biotite
	AMPHIBOLITE	K_2O+	Addition of microcline - microperthite Equigranular gneiss
			Hornblende - plagioclase ± biotite → MIGMATITE + Microcline microperthite → Plag. - qtz. - microcline - magnetite (often with destruction of gneissic structure)
Amphibolitization (a) Normal	LIMESTONE	K_2O, Al_2O_3; SiO_2; Fe_2O_3	Calc-silicate granulite : diopside - plagioclase - scapolite - microcline - biotite or hornblende - plagioclase - carbonate or biotite - plagioclase - carbonate ± garnet
(b) Alkaline → Nephelinization	LIMESTONE	Na_2O; Al_2O_3; $FeO+$	Amphibole (ferrohastingsite) - clinopyroxene (hedenbergite) - plagioclase - carbonate ± garnet ± vesuvianite ± scapolite
			Hornblende - nepheline - plagioclase - carbonate ± garnet ± scapolite ± microcline
			Clinopyroxene - nepheline - plagioclase - carbonate ± hornblende
Formation of Pyroxenite	LIMESTONE	Predominantly SiO_2+	METAMORPHIC PYROXENITE : Clinopyroxene (diopside, augite) + phlogopite ± scapolite ± titanite ± carbonate ± apatite

TABLE II

Amphibolite. As the shale becomes limy, on metamorphism hornblende appears in the assemblage. And as free silica disappears in the original sediment the equivalent metamorphic gneiss is a *biotite amphibolite* or *hornblende-biotite-plagioclase gneiss.* The typical *amphibolite* developed from a limy shale is *hornblende-plagioclase gneiss* with or without microcline, garnet, and scapolite. In the more limy rocks the assemblage may be *clinopyroxene (diopside)-hornblende-plagioclase* with or without biotite, scapolite and garnet. In the highest grades, *orthopyroxene*, usually *hypersthene*, may appear. Pyroxenic amphibolites with biotite, hornblende, and diopside are not uncommon.

The assemblages developed by regional metamorphism of limestone and sandstone require little comment here.

(ii) The following metamorphic assemblages are developed under conditions of regional metamorphism plus metasomatism:

Granitization and syenitization. In dealing with hybrid granite and syenite gneisses it is often difficult to determine what percentage of the rock is sedimentary, what part is transfused or granitized sediment, and what part is truly igneous, in that it solidified from injected magmatic material of granitic composition. Granitization and syenitization in the highland area seems to be essentially a process of potash metasomatism, and introduction of microcline and microcline perthite plays an important part.

Steps in the granitization of a paragneiss may include development of migmatite or banded gneiss, development of feldspar porphyroblasts with the wrapping around of biotite, and finally the development of an equigranular "igneous-looking" rock from the original banded gneiss.

The end product of granitization of an amphibolite is frequently the plagioclase-quartz-microcline (microperthite)-magnetite assemblage. Hornblende is replaced by magnetite, often with the destruction of the original gneissic structure of the amphibolite.

Formation of amphibolite by metasomatism. In addition to the formation of amphibolite from limy shale by a process of regional metamorphism, amphibolites are known to originate by metasomatic alteration of limestone according to the process described by Adams and Barlow (1910). Normally this process involves addition of potash, soda, alumina, silica, and iron. The result is a calc-silicate granulite with the assemblage *diopside-plagioclase-scapolite* ± *microcline* ± *biotite* or *hornblende-plagioclase-calcite* or *biotite-plagioclase-calcite.*

C. E. Tilley (Manuscript, 1954) has pointed out in his study of the Egan Chute section on the York River, Dungannon Township, that alkalic amphibolites characterized by ferrohastingsite and hedenbergite have been formed from limestones by metasomatic addition of soda, alumina, and ferrous iron. The resultant metamorphic assemblage is *hornblende (ferrohastingsite)-clinopyroxene (hedenbergite)-plagioclase-calcite* ± *garnet* ± *vesuvianite* ± *scapolite.*

As this metasomatic introduction of soda, alumina, ferrous iron, and silica proceeds to the extreme, the result is nephelinization and the formation of the assemblage *hornblende-nepheline-plagioclase-calcite ± garnet ± scapolite ± microcline* or *clinopyroxene-nepheline-plagioclase-calcite ± hornblende*. The hornblende is ferrohastingsite and the clinopyroxene, hedenbergite.

Formation of metamorphic pyroxenite. In areas such as Monteagle Township, Hastings County, where limestone bands occur as narrow remnants in a terrane of hybrid granitic gneiss, the characteristic alteration of limestone is to a granular green metamorphic pyroxenite consisting of the assemblage *clinopyroxene (augite or diopside)-phlogopite ± scapolite ± plagioclase ± apatite ± titanite.*

B. DIFFERENCES BETWEEN THE ROCKS OF THE HASTINGS BASIN AND THE HASTINGS HIGHLANDS

The essential differences between the high grade metamorphic gneiss terrane of the highlands and the low to intermediate grade rocks of the Hastings basin are summarized below.

In the Hastings basin the limestones are frequently well-bedded "dirty" or argillaceous limestones in which the impurities cannot be readily identified in hand specimen. In contrast, in the highland area the recrystallized limestones are generally coarser, and in the impure facies the impurities are present as recognizable recrystallized constituent minerals such as phlogopite, tremolite, diopside, etc. The impure limestones of the highlands are "limestones with silicates" rather than "dirty limestones."

In the Hastings basin, as exemplified by conditions in the Bancroft marble quarries, failure of the limestones may be by brecciation, while in the highland area the limestones almost invariably fail by flowage with the production of the common characteristic limestone tectonic breccias.

The argillaceous rocks of the Hastings basin as exemplified by the Hartsmere schists in Mayo Township are low grade, while their counterparts in the highland area are high grade paragneisses frequently containing sillimanite and garnet. The development of garnet as a contact metamorphic mineral in the schists adjacent to a diorite dike near Hartsmere in Mayo Township is evidence that these schists never reached the garnet grade in regional metamorphism. The diorite dike in this case also shows chilled borders which are destroyed, if ever present, in the recrystallized diorites of the highland area.

Original sedimentary features such as crossbedding and grain gradation may be preserved in the sediments of the Hastings basin. Fracture cleavage may be developed allowing structural interpretation of tops to be made. In contrast, in the highland area no original sedimentary features of use in determining tops are available. Fracture cleavage is not developed. Foliation and regional lineation may be very well developed.

Pillowed structures are frequently well preserved in the basic lavas of the Hastings basin which frequently belong to the low grade chlorite facies.

In the highland area few volcanic features are preserved and the rocks are gneiss or amphibolite of high metamorphic grade.

The diorite and gabbro intrusives of the Hastings basin are frequently fresh and show their original igneous textures. As noted above, in small dikes chilled borders are preserved. In the highland area the older gabbros have been largely converted to metagabbro, the commonest facies being a hornblende-plagioclase gneiss of granoblastic texture with clotted aggregates of recrystallized hornblende. For example, contrast the Umfraville, Thanet, and Tudor gabbro of the Hastings basin with the Mallard Lake, Raglan Hills, and Quadeville metagabbros in the highland area (Hewitt, 1953).

With respect to granitic rocks, in the Hastings basin discordant intrusives such as the McArthurs Mills granite which intrudes the crest of the Mayo anticline are present; concordant granitic rocks and "granitized" gneisses are more common in the highland area. Granites in the highland area do not show noteworthy contact metamorphic effects in the adjacent metamorphic gneisses which are already of high grade. Occasionally retrograde effects such as the development of scapolite or chlorite adjacent to intrusives are noted.

Dynamic metamorphism and "fronts." The divisions between the highlands, the Hastings basin, the Kaladar-Dalhousie trough, and the Frontenac axis are arbitrarily drawn on the basis of what the writer believes to be fairly well marked differences in lithologic assemblage, grade of metamorphism and percentage and type of plutonic rocks, including both intrusive and granitized gneisses.

It is suggested that the boundary between the Hastings basin and the Hastings highlands is marked in places by a major structural break. Within the Hastings basin the grade of metamorphism is relatively low, and sillimanite-garnet gneisses are not developed. However in Dungannon and Mayo townships as the McArthurs Mills line is approached there is a marked increase in metamorphic grade and a sillimanite-garnet gneiss zone within the Hastings basin marks what is interpreted by the writer as a frontal zone of dynamic metamorphism between the basin and the highlands.

Similarly another band of sillimanite-kyanite-garnet gneisses marks the north margin of the Kaladar-Dalhousie trough adjacent to the Fernleigh-Clyde fault, another zone of dynamic metamorphism. The zone of sillimanite gneisses developed in Hungerford, Sheffield, and Kaladar townships on the south margin of the Clare River syncline may similarly mark the southern boundary or front between the Kaladar-Dalhousie trough and the metamorphic gneiss terrane of the Frontenac axis.

A wide zone of kyanite-sillimanite-garnet gneisses extending in a northeasterly direction through Dryden Township near Sudbury may mark a zone of dynamic metamorphism near the northern border of the Grenville subprovince, the Grenville front.

2. Granites and Granitization

Within the Grenville region of central Ontario there are several different types of granite recognized as post-Grenville in age. In the area examined by the writer, these include, in probable order of age as indicated on the tentative geological timetable, grey biotite albite granite, pink microcline leucogranite, hybrid granite gneiss, pink and brown sodic granite and syenite, and biotite microcline microperthite granite.

The Boulter granite occurring in Carlow, Raglan, and Mayo townships is a large body of grey biotite albite granite. Grain size is medium; potash feldspar is present in amounts usually less than 10 per cent except when adjacent to younger pink microcline granite; there is usually a distinct foliation or lineation but inclusions or schlieren of metasedimentary material are rare. The Weslemkoon granite in Mayo and Ashby townships is a second body of this type of granite. The grey Cross Lake granite gneiss described by B. L. Smith in Clarendon and Palmerston townships is also of this general type but Smith indicates that there may be a possibility that some of this gneiss is of replacement origin.

The fine-grained pink leucogranite gneiss is exemplified by the Centre Lake and Monck Lake sheets in Cardiff Township, and the Eagles Nest, Bronson, and Great Bend bodies in Dungannon Township. It consists of an equigranular granitoid aggregate of quartz, microcline, and plagioclase with occasional biotite as an accessory. The Lavant gneiss described by B. L. Smith in Palmerston and Lavant are of this type. These rocks are similar to the Trembling Mountain gneiss of Quebec described by Osborne.

These two earlier types of granite often show a metamorphic fabric with strong regional lineation developed, and are thought to be earlier than the main period of regional metamorphism which imposed a regional lineation striking S.40°E. with a 20– to 30–degree pitch to the southeast on much of the metamorphic gneiss from Cardiff Township in Haliburton County to Lyndoch Township in Renfrew County.

Succeeding this there was a period in which much of the hybrid granite and syenite gneiss was developed: the typical Ottawa gneiss of the "fundamental Laurentian gneiss complex." Many of these hybrid gneisses are characterized by widespread development of microcline microperthite which has not been destroyed by metamorphism, suggesting that these rocks are in part post-metamorphic in age. In contrast to the Boulter and Centre Lake types described earlier, these rocks show a great variation in mineral content from place to place and much included sedimentary material.

Two types of granite found in the area are thought to be post-metamorphic. These are the dark red to yellow brown sodic syenites and granites which are often characterized by alkaline ferromagnesians such as acmite, arfvedsonite, and aegerine-augite. Microperthite is common in these rocks. The Deer Lake syenite and the Harcourt syenite in Cardiff Township and the Klondike quartz syenite in Raglan Township are of this type.

One of the youngest granites found in the area mapped by the writer is the coarse-grained pink biotite-bearing McArthurs Mills granite of Mayo Township where it occurs cutting the Hartsmere schists. The Elphin granite described by B. L. Smith in Dalhousie Township may be of this type.

Modal, chemical, and spectrographic analyses of these various types of granitic rocks, together with age determinations, are being carried out and the results will be presented in another paper.

It appears certain that granites formed both by intrusion and by replacement are present in the Grenville area. Criteria such as the presence of exsolution-type perthites as described by Tuttle (1952) in the Madoc (Deloro) granite, point to a magmatic origin. Similarly a limited mineralogic range as described by Chayes (1952) in his studies of calcalkaline New England granites also points to a magmatic origin. Evidence of processes of potash metasomatism and granitization in the Renfrew area is given by Hewitt (1953) and in the Kaladar area by Burns (1951).

Factors governing whether granite will be emplaced by intrusion or by replacement seem to be closely connected with the depth zone in which the emplacement takes place and the difference in energy level between the granitic material and the country rock at the time of emplacement.

The metamorphic environment under which the high grade metamorphic gneisses of the Hastings highland were formed must have been very favourable for granitization. This may explain the prevalence of hybrid granite gneiss in the highland area (which represents a high grade and deep zone of metamorphism), in contrast to its scarcity in the Hasting basin.

3. *Vulcanism in the Grenville*

Structural mapping of parts of the Hastings basin has shown that both acid and basic volcanics occur in the Grenville section and that these volcanics occur near the top of the Grenville section sometimes interbanded with "Hastings-type" conglomerates.

In areas of high grade metamorphism where the basic volcanics have been converted to hornblende schists and amphibolites, with the destruction of all semblance of the original volcanic structure, the interpretation of the genesis of the amphibolites is a difficult problem. As indicated in previous sections, amphibolites can also originate by regional metamorphism of a limy shale, by metasomatism of a limestone, or by metamorphism of basic intrusives. Good examples of all these types of amphibolite are available in the Bancroft area.

REFERENCES

ADAMS, F. D. and BARLOW, A. E. (1897). On the origin and relations of the Grenville and Hastings series in the Canadian Laurentian; Am. Jour. Sci., 4th Series, vol. 3, pp. 173–180.
———— (1910). Geology of the Haliburton and Bancroft areas; Geol. Surv., Canada, Memoir 6.

BAKER, M. B. (1916). The geology of Kingston and vicinity; Ont. Bureau Mines, vol. 25, pt. 3.

———— (1922). Geology and minerals of the County of Leeds; Ont. Dept. Mines, vol. 31, pt. 6.

BURNS, C. A. (1951). The Clare River area, southeastern Ontario. Unpublished M.Sc. thesis, Queen's University, Kingston.

CHAYES, F. (1952). The finer-grained calcalkaline granites of New England; Jour. Geol., vol. 60, no. 3, pp. 207–254.

CLOOS, ERNST (1934). The Loon Lake pluton; Jour. Geol., vol. 42, no. 4, pp. 393–399.

DUGAS, J. (1950). Perth map area; Geol. Surv., Canada, Paper 50–29.

ENGEL, A. E. J. and C. G. (1953). Grenville series in the northwest Adirondack Mountains; Bull. Geol. Soc. Amer., vol. 64, pp. 1013–1048.

HARDING, W. D. (1942). Geology of Kaladar and Kennebec townships; Ont. Dept. Mines, vol. 51, pt. 4.

-——— (1944). Geology of the Mattawan-Olrig area; Ont. Dept. Mines, vol. 53, pt. 6.

———— (1947). Geology of the Olden-Bedford area; Ont. Dept. Mines, vol. 56, pt. 6.

HEWITT, D. F. (1953). Geology of the Brudenell-Raglan area; Ont. Dept. Mines, vol. 62, pt. 5.

INGHAM, W. N. and KEEVIL, N. B. (1951). Radioactivity of the Bourlamaque, Elzevir and Cheddar batholiths, Canada; Bull. Geol. Soc. Amer., vol. 62, pp. 131–148.

MEEN, V. B. (1942). Geology of the Grimsthorpe-Barrie area; Ont. Dept. Mines, vol. 51, pt. 4.

MILLER, W. G. and KNIGHT, C. W. The Precambrian geology of southeastern Ontario; Ont. Bur. Mines, vol. 22, pt. 2.

OSBORNE, F. F. (1936). Intrusives of part of the Laurentian complex in Quebec; Am. Jour. Sci., vol. 32, no. 192, pp. 407–434.

PEACH, P. A. (1948). Preliminary report on the geology of Darling Township and part of Lavant Township, Lanark County; Ont. Dept. Mines, PR 1948–12.

QUINN, H. A. (1951). Renfrew map area; Geol. Surv., Canada, Paper 51–27.

SATTERLY, J. (1943). Mineral occurrences in the Haliburton area; Ont. Dept. Mines, vol. 52, pt. 2.

———— (1944). Mineral occurrences in the Renfrew area; Ont. Dept. Mines, vol. 53, pt. 3.

SMITH, B. L. (1951). Preliminary report on the geology of Clarendon Township, Frontenac County; Ont. Dept. Mines, PR 1951–3.

———— (1951). Preliminary report on the geology of South Canonto and parts of Palmerston and Lavant townships, Frontenac and Lanark counties; Ont. Dept. Mines, PR 1951–4.

———— (1954). Geology of the Clarendon-Dalhousie area; Ont. Dept. Mines, unpublished manuscript.

THOMSON, J. E. (1943). Mineral occurrences in the north Hastings area; Ont. Dept. Mines, vol. 52, pt. 3.

TILLEY, C. E. Unpublished manuscript, 1954.

TILTON, G. R. et al. (1954). The isotopic composition and distribution of lead, uranium, and thorium in a Precambrian granite; AECU-2840, U.S. Atomic Energy Commission, T.I.S.

TUTTLE, O. F. (1952). Origin of contrasting mineralogy of extrusive and plutonic salic rocks; Jour. Geol., vol. 60, pp. 107–124.

WALTON, M. S. (1953). Metamorphism and granitization in the eastern Adirondacks; abstract, Bull. Geol. Soc. Amer., vol. 64, pp. 1486–1487.

———— (1955). The emplacement of granite; Am. Jour. Sci., vol. 253, pp. 1–18.

WILSON, M. E. (1933). The Clare River syncline; Trans. Royal Soc. Can., Series III, vol. 27, Sec. 4, pp. 7–11.

———— (1940). Madoc and Marmora map sheets, nos. 559A, 560A, Geol. Surv., Canada.

WRIGHT, J. F. (1923). Brockville-Mallorytown map area; Geol. Surv., Canada, Memoir 134.

(See also Geological Survey of Canada, Reports of Progress, 1863–66; 1876–77; 1877–78, for work of Logan, Vennor, Selwyn.)

STRUCTURES IN THE CLARE RIVER SYNCLINE:

A DEMONSTRATION OF GRANITIZATION

J. W. Ambrose, F.R.S.C., and C. A. Burns

THE CLARE RIVER SYNCLINE, a well-defined fold in Grenville gneisses, lies northeast of Tweed, Ontario. Exposures in the area are good, the main features of the syncline are readily apparent, and as M. E. Wilson showed twenty years ago, structures in the granitic rocks surrounding the syncline have configurations that lead one to suspect emplacement of the "igneous" rocks was, in part if not wholly, by metasomatic processes rather than by intrusion. That is to say, the area presents structural and petrologic problems of much interest, a fact that led the late Professor E. L. Bruce to suggest to the junior author that re-study of the geology might provide material for a dissertation at Queen's University. Field work occupied two months in 1949, and four months in 1950 with a party of four. Critical places were visited by both authors. The authors are indebted to members of staff at Queen's University for many helpful discussions of problems involved. During 1949 the work was aided by a scholarship provided by the Research Council of Ontario. In 1950 the field work was supported by the Geological Survey of Canada.

The first geological report on the general area was written by W. G. Miller and C. W. Knight in 1914. In 1933, M. E. Wilson described the syncline to this Society, and in 1940 published the Madoc Sheet in which the southwestern part of the structure is displayed. In 1942, W. D. Harding published a report on the geology of Kaladar and Kennebec townships which include the northeastern extension of the structure.

STRATIGRAPHY

Wilson subdivided the surficial rocks in this area into the Grenville, as the lowermost series, overlain unconformably by the Hastings series, with the contact marked by a basal conglomerate. Earlier Miller and Knight classified certain lavas associated with the older series as Keewatin; these Wilson included with the Grenville.

As field work progressed, two, perhaps three, separate and distinct conglomerate beds were discovered within the stratigraphic column; thus the principal criterion to distinguish the base of the Hastings series could not be applied. Local names for stratigraphic units were adopted, as set out in the Table of Formations below.

TABLE OF FORMATIONS

CENOZOIC
   Pleistocene and Recent—gravel; sand, clay

PALAEOZOIC
   Ordovician

Black River Group	limestone; dolomitic limestone; arenacous limestone; sandstone

Unconformity

PRECAMBRIAN
   Pegmatite, aplite, felsite

Granitic gneisses	albite quartz monzonite; quartz monzonite; granodiorite, tonalite, diorite
Tweed Group (may be included in Kaladar Group)	crystalline limestone; biotite-amphibole schists and gneisses; conglomerate

Unconformity (?)

Kaladar Group	metavolcanics; conglomerate; biotite-amphibole schists and gneisses; crystalline limestone
Flinton Group (may be equivalent to Kaladar Group)	conglomerate; biotite-amphibole schists and gneisses; crystalline limestone, in part dolomitic and serpentinous

Unconformity

Elzevir Group	metavolcanics; amphibole schists and gneisses

The Elzevir Group is exposed only in the northwestern corner of the area. It consists of pillow lavas, amphibole schists and gneisses, and garnet-amphibole gneisses. Tops could not be determined. The group is overlain on the southeast with angular unconformity by the Flinton Group.

The Flinton Group forms a layer up to one-half mile wide that strikes a little west of south across the northwest corner of the area. It is composed mainly of conglomerate with pebbles of quartz, felsite, and granitic rocks. The remainder consists of biotite-amphibole schist and gneiss, and minor amounts of crystalline limestone.

The Kaladar Group, separated from the Flinton Group for most of its length by the Northbrook gneiss, forms a second layer across the northwest corner of the area. At its northeastern end, the group consists of a thick conglomerate bed, overlain by thin layers of biotite-amphibole schist and crystalline limestone. The conglomerate contains numerous pebbles of granitic gneiss, quartz, quartzite, and a few of amphibole-rich rocks. The group fades out gradationally northeastward into granitic gneisses.

Southwestwards the conglomerate layer is replaced by biotite-amphibole gneiss and this in turn gives way to crystalline limestone. At the western edge of the area the Kaladar and Flinton join and apparently become a single layer. This layer, shown on the Madoc Sheet, extends westwards a

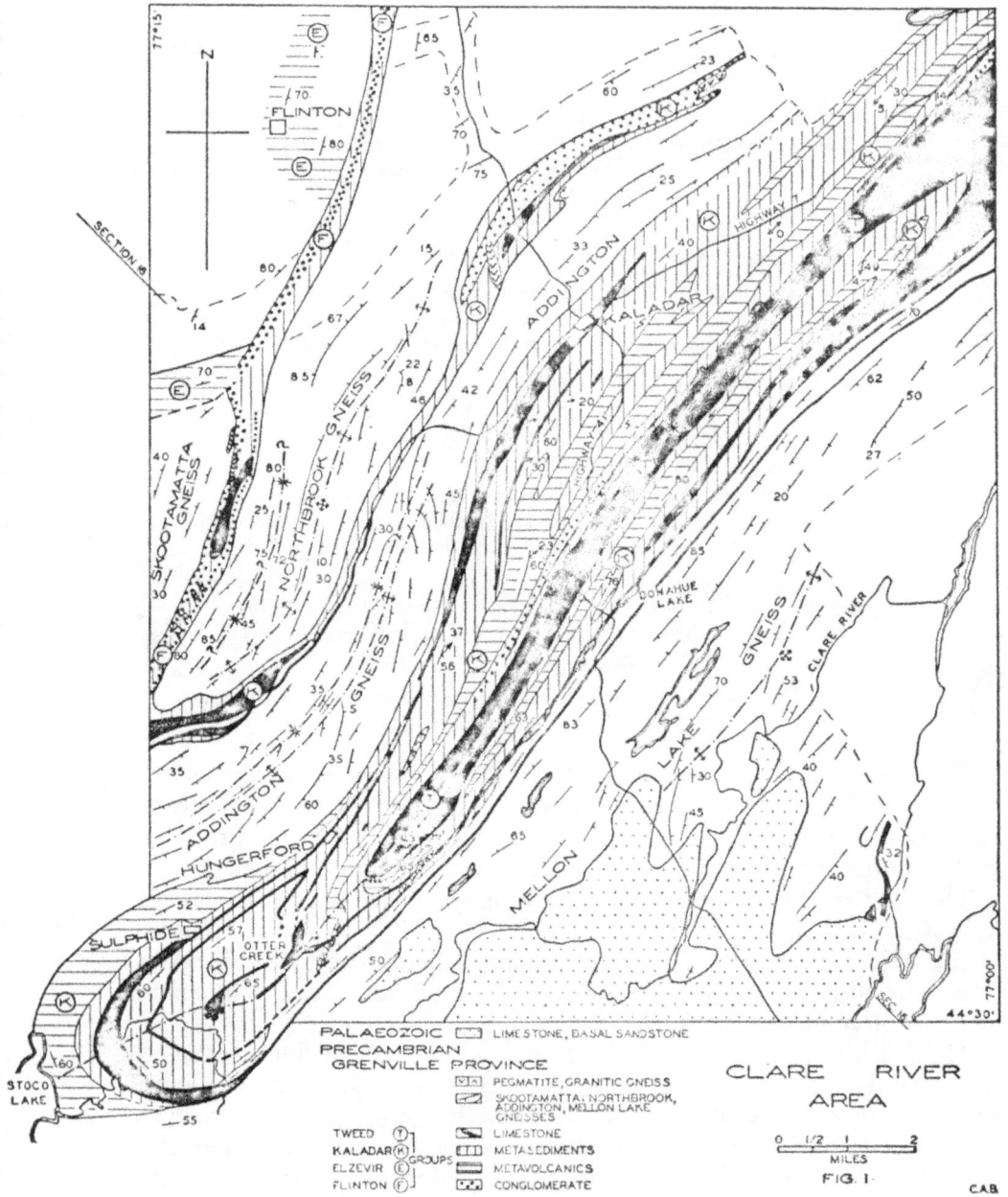

PALAEOZOIC [box] LIMESTONE, BASAL SANDSTONE
PRECAMBRIAN
GRENVILLE PROVINCE

PEGMATITE, GRANITIC GNEISS

SKOOTAMATTA: NORTHBROOK,
ADDINGTON, MELLON LAKE
GNEISSES

TWEED (T)
KALADAR (K) GROUPS
ELZEVIR (E)
FLINTON (F)

LIMESTONE

METASEDIMENTS

METAVOLCANICS

CONGLOMERATE

CLARE RIVER
AREA

0   1/2   1   2
MILES

FIG. 1

C.A.B.

FIGURE 1. Clare River area.

further twelve miles and develops at least three remarkable, hook-like structures along its southern flank. Were it not for this unexplained, and as yet inexplicable, complication one would have no hesitation in correlating the Flinton and the Kaladar groups, and thus establishing the structure they outline as a syncline plunging northeastwards.

Structures in the intervening Northbrook gneiss show reversals of dip which can be interpreted as a syncline with a small flanking anticline; this supports the view that the Flinton and the Kaladar are correlative. However, the attitudes can also be interpreted as an anticline with a small flanking, perhaps even questionable syncline (Fig. 2). Northeastwards the attitudes of gneissosity become parallel and continuity of the structures is lost. Final decision as to the relations between the two groups, and the structure, must depend on re-study of their western extension.

This layer of the Kaladar Group is separated from the main Clare River syncline by a layer of gneiss one to two miles thick, shown in Figure 1 as the Addington gneiss. Planes of gneissosity are strictly conformable with the sedimentary contacts. Along and northeast of Highway 41 the gneissosity dips consistently 50 to 70 degrees southeast, but southwest of the highway the area of gneiss widens and foliation planes here as in the Northbrook gneiss show reversals in dip which can be interpreted as an anticline with a small flanking syncline, passing northeastwards into an isoclinal fold. Here again, as in the area between the Flinton and the lower Kaladar, the attitudes in this gneiss layer may signify nothing more than two small complementary wrinkles. With this interpretation, illustrated in Fig. 2, the gneissic layer between the lower Kaladar and the main body of the syncline occupies the position of a simple stratigraphic unit.

Interpretation of relations between the Kaladar Group and the rocks that form the main body of the syncline depends largely on interpretation of the structural position of the Addington gneiss, and this in turn depends on interpretation of the Flinton-Kaladar relations and structure. Evidently an understanding of the geology of the western extension is of critical importance for any clear understanding of the geology of this area.

The main Clare River syncline occupies a diagonal strip two to five miles wide which extends across the area from the southwest corner to and several miles beyond the northeastern boundary. (See also Harding-Kennebec and Kaladar Sheets.) The rocks within the syncline consist of a lower series of pillow lavas and metavolcanics, exposed around the nose and along the northwest flank to Hungerford. These are overlain by an alternating succession of thick biotite-amphibole gneisses, some garnet-sillimanite gneiss, and thin but remarkably persistent layers of white to grey, coarsely crystalline limestone which serve as marker beds. This alternating series is overlain by a second thick layer of metavolcanics, here taken as the uppermost part of the Kaladar. These metavolcanics, although more variable than either those at Hungerford or in the Elzevir Group, have compositions and structures which demonstrate their volcanic origin. Thus three periods of

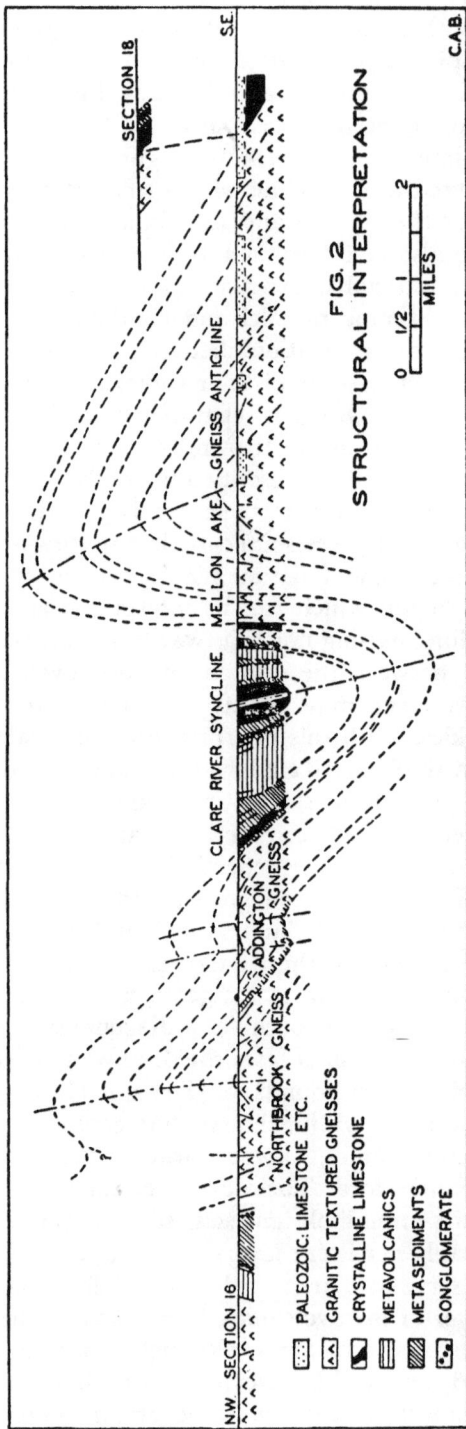

FIGURE 2. Structural interpretation.

volcanism are indicated, and period groupings based on the presence or absence of volcanic rocks are of little use in this area.

This section, from the base of the lower metavolcanics in the Clare River syncline to the top of the upper metavolcanics, was grouped with the Kaladar. This was done because the Addington gneiss was regarded as a stratigraphic unit, not a fold, which separates a lower, conglomerate-bearing member as described above, from the main synclinal sedimentary rocks. However, because of the difficulties of interpretation of the structures in, and the structural position of the Addington gneiss, and because of the uncertain relations between the Kaladar and the Flinton, it seems best to reserve judgment until the western area is studied.

The upper metavolcanic series is overlain by another sedimentary series, the Tweed Group. The lower part of this group consists mainly of biotite-amphibole gneiss, but includes, north of the synclinal axis, a thin conglomerate member traced for four miles near Highway 41, and another thicker member near the eastern boundary. These clastics are overlain by thick crystalline limestones which occupy the trough of the syncline. The presence of the conglomerate overlying a volcanic series and grading upwards into finer clastics seems to indicate an erosional interval. However, no structural discordances were noted and the unconformity may not have great significance.

Along the southeastern flank of the syncline biotite-amphibole gneisses and crystalline limestones are in contact with conformable granitic gneisses, the Mellon Lake gneisses. Foliation planes in these gneisses outline a broad anticline with its axial plane dipping steeply southeast. The remnants of limestone in the extreme southeastern corner of the area lie on the southeastern flank of the fold and might therefore be correlative with limestones in the main syncline.

The petrography of the granitic gneisses deserves a word. The rocks are all coarse to medium grained, pink to grey weathering, with foliation planes marked by aligned biotite, amphibole or feldspar crystals, or by thin blades of quartz. They range in composition from tonalite through granodiorite to quartz monzonite. True potash-rich granites are absent.

Correlations can be made between some of the areas of granitic gneisses. The Mellon Lake gneiss is continuous around the southwest end of the Clare River syncline, and attitudes shown in the Madoc Sheet indicate that the synclinal structure persists southwestwards, in the gneisses, at least for a short distance. That is, the Mellon Lake gneiss anticline appears to pass northwestward into a syncline wrapping around the Clare River syncline, and the Mellon Lake gneiss is in part at least one and the same as the Addington gneiss. Since the Addington gneiss and the Northbrook gneiss become one in the northeastern part of the area there may be only one gneiss in the whole area. However such generalizations are unsafe as yet and must wait until more detailed studies can be made of the interrelations of the several gneiss facies.

Structure

Clearly, interpretation of the stratigraphic succession depends heavily on structural interpretation. Exposures within the area (Fig. 1) are excellent, limestones and conglomerates form easily traceable marker beds, and in one or two places beds of garnet-sillimanite gneiss can be followed for some distance. No markers were discovered within the granitic gneisses. Bedding is well displayed in the sedimentary strata, as are gneissosities in the granitic rocks. Tops were determined in only one place, for although several outcrops of pillow lavas were found, only those at Sulphide could be interpreted. There the flows face north, away from, instead of in towards, the synclinal axis. This anomalous position appears to be due to a small local fold. Small drag folds occur at several places but their usefulness is limited because the beds in which they occur, mainly limestones, are sharply plicated into folds of several orders of size. Plunges of drag folds probably reflect the plunge of the major folds.

The structure is illustrated in Figure 1 and a tentative interpretation is offered in Figure 2. The southwest end of the Clare River syncline, outlined by metasediments and metavolcanics, outcrops northeast of Stoco. Beds in the southwesterly part of the syncline and on most of the southern limb dip steeply northwest; those on the northern limb north and east of Kaladar are less steeply inclined to the southeast. That is to say, northeast from Kaladar where the syncline is widest—five or six miles across—the axial plane dips southeast and the fold is inclined slightly to the north.

Plunges of crenulations and dragfolds, at and near the nose of the major syncline, are easterly at 35 to 50 degrees. Northeasterly the plunge decreases; east of Kaladar it is horizontal, east of there the plunge is gently southwest; that part of the syncline in this area is evidently deepest where it is widest.

Three small structures that illustrate the degree of compression and contortion are exposed in the southwestern part of the fold (Fig. 3). One of these probably involves the pillow lavas at Sulphide and accounts for the fact that these flows apparently face in the "wrong" direction.

A second small structure exposed southwest of Hungerford is depicted by the convergence northeastwards of two limestone layers. Two other, lower, limestone members converge and join northeast of Hungerford. The single layer thus formed extends for a further half-mile, where it fingers and slivers out. The inference that the structure is a fold is supported by the presence of a third marker, a layer of garnet-sillimanite gneiss exposed north of the north limestone layer and south of the southern. This part of the fold is similar to but is more closely compressed than that southeast of Hungerford.

A third minor complication is exposed north of Otter Creek. Here the form of the syncline is outlined by a continuous uncontorted bed of limestone, but a second layer higher in the succession is sharply folded and crenulated. Folds of several orders of size are developed; many show

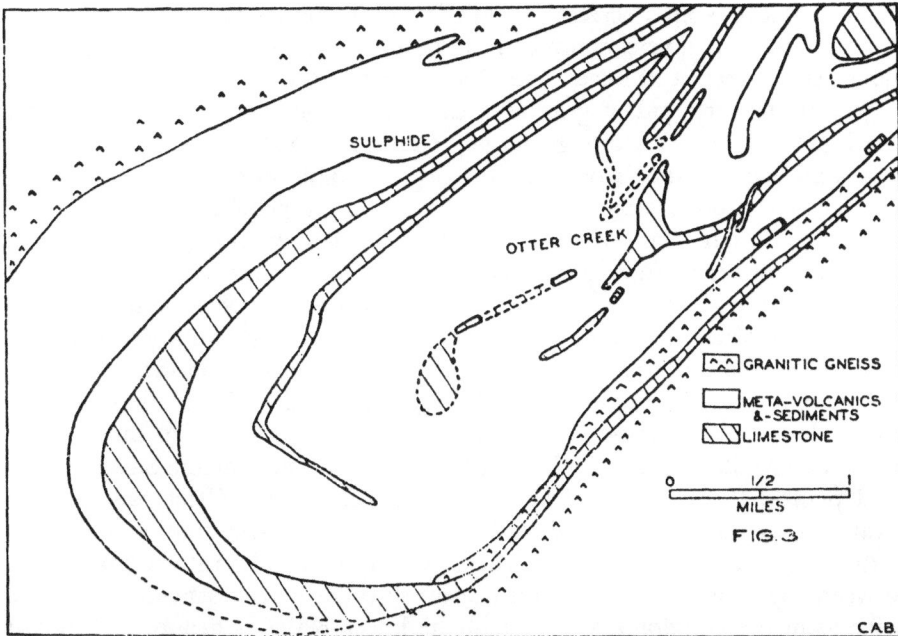

FIGURE 3. Compression and contraction, Clare River area.

thickened crests and thinned limbs, and correlations of single beds from outcrop to outcrop are difficult or impossible. In spite of their complicated forms these small folds all plunge northeast, as is to be expected in this part of the syncline.

Thus the Clare River syncline proves to be a fairly simple canoe-shaped fold, overturned slightly towards the northwest, and with a few local, relatively minor complications. Faults of any consequence appear to be absent.

STRUCTURE AND GRANITIZATION

The structural situation along the southeastern flank of the syncline bears directly on one of the problems in hand, i.e., the manner of emplacement of the granitic gneisses. There a layer of crystalline limestone, and skarn derived from carbonate-rich rocks, is separated from the main sedimentary series by a layer of granitic gneiss 200 to 300 feet thick. The carbonate-rich layer can be traced through intermittent outcrops for nearly fifteen miles; throughout this distance it is parallel to the main contact and shows no disruption or deviation whatever. When this layer is followed southeast it swings into and becomes continuous with the outermost limestone layer in the Kaladar series.

An almost identical situation is developed along the northern flank northeast of Kaladar. Here the lower limestone layer can be followed through granitic gneisses for at least four miles. Over this distance the biotite-amphibole gneisses which lie above and below the layer have disappeared and their place has been taken, as along the southern flank, by layers of granitic gneiss.

The situation seems perfectly clear. The sedimentary layers on either side of these limestone layers, composed mostly of mixtures of biotite, amphibole, oligoclase, quartz, and accessories, have been selectively replaced by rocks of granitic texture and of granodioritic to quartz-monzonitic compositions. The limestone layers, although largely converted to skarn, resisted replacement and retain enough of their identity to enable one to recognize them without hesitation. It seems to the authors to be impossible to reconcile the existence of such thin, weak, partitions for miles within granitic gneisses with any process of forceful intrusion, either by stoping or by shouldering aside. They see no escape from the conclusion that here at least the granitic gneisses developed by some process of quiet (as opposed to disruptive) volume for volume replacement, that is to say, by metasomatism.

This conclusion is strongly supported by strict conformity of the structures within the granitic gneisses with those that can be observed in the surficial rocks. The Mellon Lake anticline and the Clare River syncline have parallel axial planes; the folds are simple and complementary. Strict concordance obtains everywhere between attitudes of gneissosity in the granitic and those in the sedimentary and volcanic rocks. Where surficial strata disappear, as between the limestone layers and the other rocks of the syncline, or along strike as in the conglomerate layer of the lower Kaladar group, or through lit-par-lit injection, as along Highway 41 southeast of the main syncline, the attitudes are constant, undisturbed, and undistorted. Attitudes of and within lenses of granitic gneiss in the main syncline and within xenoliths in the granitic gneisses are strictly concordant.

It seems impossible to reconcile these facts with any theory of emplacement of the "igneous" rocks other than by metasomatic replacement. Stoping is clearly out of the question unless one is prepared to believe that such a process is capable of precise selection, which seems absurd. Similarly all the evidence argues against any process of shouldering aside, or even of permissive occupation of opening spaces. On the contrary, pseudomorphism of earlier structures seems complete in every detail.

One other possibility remains. A. F. Buddington[1] suggests that certain granitic phacoliths in the Adirondacks were intruded as sills into a series of flat-lying stratified rocks and that the whole was then folded into its present configuration. This process, as the authors understand Buddington's description, requires extensive and rather complete mylonitization of the intrusive rocks, followed by folding and subsequent re-crystallization. In the Clare River area the surficial rocks and perhaps the granitic gneisses are high rank metamorphic rocks and as such have been completely re-crystallized. The Mellon Lake gneiss in the northeastern part of the area has a partly cataclastic (protoclastic?) texture. Micropherthite, which Buddington finds unmixed and which thus is absent in the re-crystallized rocks of the phacoliths, occurs only rarely in these gneisses. However, the preservation of recognizable pillows, the preservation of sharply contrasted compositional layering (as between thin limestone layers and layers of

[1]A. F. Buddington, oral presentation, Royal Society of Canada, June, 1955.

biotite-amphibole gneiss), and again, the existence of thin layers of lime-stone structurally undisturbed in relation to the main fold axis, argue strongly against the possibility that such a process may have operated here. If cubic miles of intrusive rock were mechanically reduced to a mass sufficiently mobile to form folds and develop concordant gneissosity, the limestone partitions would have been affected in the same way as are layers of paragneiss and quartzite in folded crystalline limestones, that is to say, they would surely have been disrupted into a disjointed tectonic breccia, or perhaps would have flowed into forms like those described by D. F. Hewitt from Haliburton County.[2] This is not intended as a summary dismissal of an hypothesis which explains many remarkable and interesting structures in the Adirondacks. In the Clare River area, on the other hand, the evidence presently assembled seems to weigh rather conclusively against its acceptance as a valid explanation.

A word about the minor dykes is necessary. Such dykes are uncommon in the area generally, and most of them are concordant. However, a few do cross-cut the strike in places, for example, in the southwestern part of the Flinton Group. These bodies were not studied in sufficient detail to distinguish them either as dilational or as replacement dykes.

Emplacement of the Granitic Gneisses

Pertinent evidence concerning emplacement of the granitic gneisses can now be summarized as follows:

1. The types of rocks involved include volcanics, limestone, calcareous amphibole rocks, alumina-rich metasediments, feldspathic gneisses all of surficial origin and granitic-textured gneisses of igneous appearance which range from albite quartz monzonite through quartz monzonite and granodiorite to tonalite and diorite.

2. Configuration of gneissic foliation planes indicates that structures are continuous from metasediments and metavolcanics into and through large areas of, and thus large "stratigraphic" thicknesses of, granitic gneisses.

3. Thin layers of limestone lie unbroken and continuous within gneisses for as much as fifteen miles. At least two layers join with and form part of the metasediments of the main syncline.

4. All schlieren are concordant with the enclosing granitic gneisses, and thus with the adjacent sedimentary and volcanic structures.

5. Within the main syncline, lenses of granitic gneisses are interlayered with the metasediments like stratigraphic units.

6. In a few places in the northwestern part of the Mellon Lake gneiss, reversals in dip and local irregularities occur associated with many pegmatitic dykelets and quartz stringers. Here the texture of the gneisses is in part cataclastic, perhaps even protoclastic, and the order of crystallization is difficult to establish. Aside from this there is no evidence to indicate that the granitic gneisses were disturbed after they were formed other than by regional uplift.

[2]Oral presentation, Royal Society of Canada, June 1955.

7. Gradation from metasediments to granitic gneisses is evident in several places. On the south side of the syncline, southwest of Highway 41, calcareous biotite-amphibole gneiss grades into granitic gneiss partly through *lit-par-lit* mixtures and partly by an abrupt gradation. In the same general area a layer of sillimanite garnet gneiss resembles, megascopically, granitic gneiss but contains a little sillimanite and garnet perhaps of sedimentary origin. North of Kaladar, conglomerate can be traced to outcrops where the pebbles are ribbons, and the matrix is similar to granitic gneiss. The rocks grade into one another and the boundary between conglomerate and granitic gneiss is arbitrary.

Thus in this area no item of evidence supports an hypothesis of emplacement of the granitic rocks either by forcible shouldering aside of the country rock, or by engulfment, either piecemeal or in large blocks. Instead all the evidence seems to weigh heavily in favour of a process of quiet development of rocks of igneous appearance from rocks already there. In short, the emplacement must have been by molecular replacement, the gneissosity represents relict bedding, and the area constitutes a demonstration of metasomatic granitization. In one place only, in a small area within the Mellon Lake gneisses, did the process become sufficiently advanced to permit even partial mobility of the rocks in the magmatic sense.

The statement that this constitutes a demonstration of metasomatic granitization is by no means intended to imply that the authors are prepared to offer an explanation of the process. The investigation which any such attempt would require is beyond the scope of this paper. It is interesting to note that compositions of the gneisses include rocks classifiable as quartz monzonite, granodiorite, and tonalite and include only a very few, if any true, potash-rich granites. Explanations for this, and for the many problems associated here as elsewhere with metasomatic granitization, must await much additional information and study.

Finally, the stratigraphic succession in this area is only partially deciphered. One unconformable contact seems established, a second may be present, three periods of volcanism are recognized, correlations of some of the sedimentary groups with one another and of some layers of granitic gneiss with one another seem reasonable. Much remains to be learned by critical re-study, in particular of the westward extensions of the Flinton and Kaladar groups. The Grenville-Hastings relationship would form a fundamental problem in this larger study.

REFERENCES

HARDING, W. D. (1942). Geology of Kaladar and Kennebec townships; Rept. Ont. Dept. Mines, vol. 51, pt. 4.

McKINSTRY, H. E. (1948). Mining geology. New York: Prentice-Hall.

MILLER, W. G. and KNIGHT, C. W. (1914). Geology of southeastern Ontario. Rept. Ont. Dept. Mines, vol. 22, pt. 2.

WILSON, M. E. (1933). The Clare River syncline; Trans. Roy. Soc. Canada, Series III, vol. 27, Sec. IV, pp. 7–11.

———— (1940). Madoc Map Sheet, Ontario; Geol. Surv., Canada, Map. 559A.

DISCUSSION

M. E. WILSON

It is now thirty years since the field work on the Madoc and Marmora map areas was completed. Although at that time I noted that an important un-conformity separates the Hastings series from the pre-Hastings, presumably, Grenville rocks, all of the granitic rocks appeared to intrude the Hastings sediments, and I could find no good evidence that mountain building had intervened between the two series. Since then, however, a band of conglomerate containing abundant boulders of granite, has been discovered crossing Highway 41, about two miles north of Kaladar. This indicates that a Hastings-Grenville unconformity is structural as well as erosional.

On the map of the Madoc area, the area of greywacke and limestone that lies between the conglomerate of the Clare River syncline and the underlying volcanic rocks to the north, was shown as Grenville series. This may have been a mistake. It is possible for pebbly beds to be deposited above mud on a beach, and therefore that the pre-Hastings unconformable contact underlies the grey-wacke. If this is so, then all the volcanic rocks in the Madoc and Marmora map areas are in the Grenville series and all the sediments, except for some dolomite and quartzite, belong to the Hastings series. Furthermore, the blue-weathering limestone that occurs so extensively in the Madoc-Marmora and Bancroft map areas overlies the Hastings conglomerate and is part of the Hastings series. This suggests the possibility that part, at least, of the crystalline limestone in the Grenville subprovince, is Hastings limestone recrystallized.

The pillow-like forms in metamorphosed volcanic rock at Sulphide de-scribed by Burns lie within a special map that I made of an area adjacent to the Sulphide pyrite mine on the scale of 500 feet to 1 inch. Recently I re-examined the outcrops in this area, and observed dark grey crumpled zones about one-quarter of an inch wide in a fine-grained phase of lava which possibly outline pillows. Only three bun-like forms possibly indicate tops; two to the north and one to the south were seen. These can scarcely be used with certainty to determine tops. However, if mountain building occurred in the interval between the deposition of the Grenville lavas and of the Hastings series, the possible existence of pillow with tops north at Sulphide, would be further proof of structural unconformity, but without structural significance within the Clare River syncline.

The occurrence of thin bands of limestone in gneissic granite on the south side of the Clare River syncline presents a difficult problem. In Quebec, north of Ottawa, immense areas of limestone or dolomite have been transformed to diopside by siliceous emanations from granite and pegmatite, and the grey-wacke of the south limb of the Clare River syncline about 1000 feet to the north has been lenticularly silicified. If the granite enclosing the limestone is transformed greywacke, how did the limestone remain unaltered while this transformation was taking place? Probably a better explanation of the relation-ships of the Clare River syncline to the enclosing gneissic granite is that the granite while in a plastic condition underwent folding with the sediments.

THE BEARING OF AGE DETERMINATION

ON THE RELATION BETWEEN THE

KEEWATIN AND GRENVILLE PROVINCES

H. A. Shillibeer and G. L. Cumming

IN THIS PAPER the authors present a summary of the age determinations on rocks and minerals from the Keewatin and Grenville geological provinces of the Canadian Precambrian. There are now available over 140 age determinations from these two regions, about 100 of which are from the Grenville province.

The majority of the analyses used for these age determinations have been made on minerals from pegmatite dykes and vein deposits and hence give a lower limit to the age of the surrounding rocks, but it has always been supposed that pegmatites are only slightly younger than the igneous rocks which they accompany. This has been shown to be true in the case of the Essonville granite (Tilton *et al.*, in press). It has always been considered that granites are intruded at the time of maximum mountain building disturbance. Thus by dating pegmatites it is possible to study what is considered to be the primary problem in the Precambrian, that is the distinction between regions of separate mountain building activity.

It has also been suggested by Wilson (1954) that in any region the age of pegmatites, and the orogenic activity which produced them, is within a very few hundred million years of the age of the sediments and volcanic rocks, and this is certainly true in younger mountain systems. For example, Rodgers (1951) has summarized the available data for the Appalachian region, and the range of ages of pegmatites (700 to 200 million years) corresponds to the time when the sedimentary and volcanic rocks of this region were being formed. Recent isotopic age determinations (Table I) on Appalachian pegmatites corroberate the data collected by Rodgers.

Since it is generally agreed that the Grenville province represents the eroded roots of an old mountain system, it is reasonable to suppose that the same relations between igneous and sedimentary rocks apply to this region and to the Keewatin province as well. Proof of this hypothesis, which depends on dating both sedimentary and igneous rocks, is not complete for Precambrian areas but it is irrelevant to the more general problem of determining the periods of orogenic activity.

The various methods of age determination have been thoroughly dis-

TABLE I

ISOTOPIC AGES FROM APPALACHIAN MINERALS

URANIUM-THORIUM MINERALS

Location	$\dfrac{207}{206}$	Age m.y. $\dfrac{206}{238}$	$\dfrac{207}{235}$	$\dfrac{208}{232}$	Analyst
Mitchel Co., N.C.	355 ±40				Collins (1954)
Spruce Pine, Avery Co., N.C.	699	312	362	354	Kulp (1953)
Branchville, Conn.		367	365	318	Wasserburg (Program Abstracts of the A.G.U., May, 1955)

POTASSIUM MINERALS

Location	$\dfrac{A^{40}}{K^{40}}$	Age m.y.	Analyst
Branchville, Conn.	0.0168	353	Wasserburg (*Ibid.*)
Bedford, N.Y.	0.0177	370	Wasserburg (*Ibid.*)
Spruce Pine, N.C.	0.0160	330	Shillibeer (1955)

cussed by many authors (Nier, 1939; Collins *et al.*, 1954; Kulp *et al.*, 1954; Faul, 1954; and Wilson *et al.*, in press). For the purpose of this paper it will be sufficient to point out that isotopic analyses are essential for first-class determinations, and that only the uranium-lead and thorium-lead methods are firmly established. Although the potassium-argon method is being developed as an independent technique, it still ultimately relies on calibration against the uranium-lead methods (Shillibeer *et al.*, 1954). The ages obtained from isotopic analyses of lead minerals are based on calibration against uranium-lead ages (Russell *et al.*, 1954), but where comparisons are possible the results obtained are often in striking agreement with ages obtained by other methods. The results quoted in the tables by the rubidium-strontium and helium methods, though less reliable, may be taken as rough comparison ages where no other data are available.

In interpreting a particular age determination, it is important to consider the source of the material, whether vein, pegmatite, or rock, as well as the geological province in which it occurs. In Table II (printed at the end of the paper) the age determinations from the Grenville and Keewatin provinces are given. It is apparent that pegmatites from the Keewatin region are all about 2500 million years old and hence the latest orogenic activity took place about that time. The majority of the galenas are also about this age, particularly those from the gold deposits of Ontario and Quebec. A few galenas from the Keewatin province yield younger ages. Some of these may be anomalous in the sense used by Russell *et al.* (1954), and the remainder are probably minor deposits of younger age. The helium ages of Hurley (1949) on magnetite from Keewatin rocks also indicate ages of greater than 2000 million years. Thus, age determinations from three quite different sources, pegmatites, veins, and rocks, all indicate that the major orogenic activity in the Keewatin region took place between 2000 and 2500 million years ago and that following this activity the greater part of this region has been dormant, since only a few vein minerals give ages younger than this and they were probably introduced as fracture fillings.

A large group of isotopic ratios have been published (Russell *et al.*, 1954 and Cumming *et al.*, 1955) for lead ores from the Sudbury-Cobalt region. This region is underlain by basement rocks of the Keewatin province. However the ore deposits are related to intrusives which cut relatively unaltered Proterozoic-type sediments. An age of about 1200 to 1300 million years has been assigned to the Sudbury ores (Russell *et al.*, 1954) indicating that the Huronian sediments are older than this age. On the other hand, many geologists consider that the Sudbury intrusive, to which the lead ores are closely related, was emplaced under the Whitewater series while they were still unconsolidated. Therefore, the time of deposition of these sediments may not be much greater than the age of the intrusive and its associated ores.

The Cobalt ores are also considered to be anomalous. The great variations in their isotopic constitutions are considered to be due to radiogenic additions to Keewatin leads which were reworked and redeposited at the same time as the Sudbury ores (Cumming, 1955).

Another group of age determinations from veins in the Keewatin region, close to its southeastern boundary, is that from the Theano Point district. These deposits are closely related to late diabase dykes and their age is about 1100 million years.

There are thus several localities in the Keewatin province, along its boundary with the Grenville province, where vein minerals were deposited about 1100 to 1300 million years ago. The age determinations on these ores suggest that the Huronian rocks which overlie the Keewatin basement are closely related in time to the Grenville mountain system, and Wilson (1954) has suggested that they bear the same physical relationship to the old Grenville mountains as do the Rocky Mountains to the Coast Ranges of the western Cordillera. That is, the lower Huronian rocks are the same as some of the Grenville sediments, while the upper Huronian rocks were derived from the Grenville mountains, and laid down in basins adjacent to the mountains on a pre-existing basement and on remnants of the lower Huronian.

The Grenville region has been well dated only in the southwestern part and here it seems clearly established that granitic rocks and pegmatites are all between 800 and perhaps 1350 million years old. The only pegmatite minerals which indicate ages younger than this range are a few rare earth silicates and thucolites. In some cases these occur in dykes which have been dated by other minerals at about 1000 million years and the evidence for partial alteration of the dyke at a later time seems clear.

The most significant fact, however, is that no ages are known in the Grenville province which are older than 1350 million years. If any older igneous rocks exist in this region, it is indeed peculiar that none have so far been discovered. This upper limit of about 1350 million years has been corroborated by one of the authors for a suite of igneous rocks and associated aplites and pegmatites in the Wassau region of Wisconsin where the ages obtained range from 1100 to 1350 million years (Shillibeer, 1955).

It is thus quite clear that the major igneous activity, the formation of metamorphic rocks and the intrusion of pegmatites in the Keewatin and Grenville regions, are totally unrelated in time, and took place about 2500 million years ago and between 1300 and 800 million years ago respectively.

Conclusion

In younger mountain systems, the period of igneous activity closely parallels the time of formation of sedimentary and volcanic rocks, and hence it seems reasonable to conclude that this parallelism existed in older mountain systems as well.

If this is so, then the conclusion that the continents have grown by successive marginal accretion seems inescapable.

Age determinations indicate that the Keewatin and Grenville provinces of the Canadian Precambrian are distinct orogenic units which are unrelated to one another and which represent two successive periods of the growth of the continental mass during geologic time.

Acknowledgments

The authors wish to thank Dr. J. T. Wilson for his continued interest and many useful suggestions, and Dr. R. D. Russell and Dr. R. M. Farquhar for permission to include their unpublished isotopic analyses of galenas.

REFERENCES

Collins, C. B., Farquhar, R. M. and Russell, R. D. (1954). Bull. Geol. Soc. Amer., vol. 65, p. 1.

Cumming, G. L. (1955). Unpublished Ph.D. thesis, University of Toronto.

Cumming, G. L., Wilson, J. T., Farquhar, R. M. and Russell, R. D. (1955). Proc. Geol. Assoc. Can., vol. 7, pt. 2, p. 27.

Faul, H., ed. (1954). Nuclear Geology. New York: John Wiley and Sons.

Hurley, P. M. (1949). Science, vol. 110, no. 2845, p. 49.

Kulp, J. L., Eckelmann, W. R., Owen, H. R. and Bate, G. L. (1953). U.S. Atomic Energy Comm. Rept. NYO–6199.

Kulp, J. L., Bate, G. L. and Broecker, W. S. (1954). Am. Jour. Sci., vol. 252, p. 345.

Nier, A. O. (1939). Phys. Rev., vol. 55, no. 2, p. 153.

Rodgers, J. (1951). Bull. Geol. Soc. Amer., vol. 62, p. 1561.

Russell, R. D., Farquhar, R. M., Cumming, G. L., and Wilson, J. T. (1954). Trans. Am. Geophys. Un., vol. 35, p. 301.

Shillibeer, H. A. (1955). Unpublished Ph.D. thesis, University of Toronto.

Shillibeer, H. A. and Russell, R. D. (1954). Can. Jour. Phys., vol. 32, p. 681.

Tilton, G. R., Patterson, C. C., Brown, H., Inghram, M., Hayden, R., Hess, D. and Larsen, E. S. (in press). Bull. Geol. Assoc. Amer.

Wilson, J. T. (1954). The Earth as a Planet, ed. G. P. Kuiper, chap. 4. University of Chicago Press.

Wilson, J. T., Russell, R. D. and Farquhar, R. M. (in press). Handbuch der Physik, vol. 47, ed. J. Bartels, chap. 14. Berlin: Springer-Verlag.

TABLE II

Deposits of Grenville Province

Location and mineral[1]	Analyst	U(%)	Th(%)	Pb(%)	204	206	207	208	Age[2]	Reference[3]
Anacon Mine 423 stope, P.Q.	G. Tor. #559[4]				1.436 / 1.00	23.76 / 16.55	22.46 / 15.64	52.36 / 36.46	1240 ±190	(f) Proc. Geol. Assoc. Can. 7: Pt. 2, 27, 1955
Anacon Mine 615-60 stope, P.Q.	G. Tor. #560				1.437 / 1.00	23.75 / 16.53	22.46 / 15.64	52.34 / 36.44	1240 ±190	(f) Ibid.
Anacon Mine 826-50 stope, P.Q.	G. Tor. #561				1.441 / 1.00	23.76 / 16.49	22.45 / 15.58	52.35 / 36.33	1285 ±190	(f) Ibid.
Anacon Mine P.Q.	G. Tor. #408	Identical material to Nier below			1.426 / 1.00	23.98 / 16.82	22.39 / 15.70	52.22 / 36.62	1115 ±195	(f) Ibid.
Anacon Mine (Tetrault) P.Q.	G. Nier				1.470 / 1.00	23.92 / 16.27	22.28 / 15.16	52.33 / 35.6		Phys. Rev. 60: 112, 1941
Bathurst Twp. Ont.	U. Tor. #57				0.014	89.43	6.85	3.73	1080 ±30	(a)
Besner Mine, Henvey Twp., Ont.	U. Ellsworth	68.29 / 67.26 / 67.30	1.56 / 1.52 / 1.85	8.04 / 7.51 / 7.57	heavy conc. (best material) / light conc. single crystal					Am. Min. 16: 576, 1931
Besner Mine	U. Nier	Used average of Ellsworth's analyses / Recalculated using best analysis of Ellsworth			0.0068 / 0.0068	92.77 / 92.76	6.24 / 6.24	0.98 / 0.99	825 / 770 / 785 / 845	(a) (b) (c) (d) Phys. Rev. 55: 158, 1939
Besner Mine	Cy. Muench	1.83	0.01	0.036					140	(e) J.A.C.S. 58: 2433, 1936
Besner Mine	U. Tor. #299[4]				0.02[1] ±0.007	91.86	6.71	1.44	940 ±80	(a) Proc. Geol. Assoc. Can. 7: Pt. 2, 27, 1955

This sample may not be from Besner Mine. Cf. Tor. #229 which was obtained from the Besner workings.

TABLE II (cont'd)

Location and mineral[1]	Analyst	U(%)	Th(%)	Pb(%)	204	206	207	208	Age[2]	Reference[3]
Besner Mine	U. Tor. #423	Identical lead to Nier. Ages calculated using Ellsworth analysis			0.011	92.38	6.41	1.20	860 (a) / 770 (b) / 800 (c) / 845 (d)	Ibid.
Besner Mine	Th. Muench	4.63 / 1.24	0.903 / 4.21	0.186 / —	nodular material pseudocrystalline				270 (e)	J.A.C.S. 59: 2269, 1937
Besner Mine	Th. Nier	Same material as Muench			0.021	88.65	5.21	6.12	440 (a) / 262 (b) / 275 (c) / 250 (d)	Phys. Rev. 60: 112, 1941
Besner Mine	F. Tor. #1178A	%K$_2$O = 13.2 ± 0.2 (Watson)[5]			$\frac{A^{40}}{K^{40}} = 0.052$				900 ± 70 (g)	Proc. Geol. Assoc. Can. 7: Pt. 2, 27, 1955
Besner Mine	F. Wasserburg & Hayden	%K$_2$O = 10.98			$\frac{A^{40}}{K^{40}} = 0.0542$[5]				925 (g)[6]	Phys. Rev. 93: 645, 1954
Butt Twp. Ont.	U. Ellsworth	64.23 / 66.01 / 65.19 / 67.40	0.99 / 1.08 / 1.61 / 1.37	9.61 / 9.82 / 9.13 / 10.02	best crystal of the four					G.S.C. Econ. Geol. Ser. #11, 1932
Butt Twp. Ont.	U. Tor. #380	Same material as Ellsworth			0.007	92.23	6.79	0.98	1020 (a) / 960 (b) / 980 (c) / 1150 (d)	
Cardiff Twp.	Ell. Ellsworth	18.39	0.09	1.61					630 (e)	G.S.C. Econ. Geol. Ser. #11, 1932
Cardiff Twp. Ont. (Monck Lake granite)	Gr. Tor. #1274A	%K$_2$O = 3.70 (Moddle)[5]			$\frac{A^{40}}{K^{40}} = 0.047$				830 ± 70 (g)	

Footnotes at end of Table.

TABLE II
Deposits of Grenville Province (cont'd)

Location and mineral[1]		Analyst	U(%)	Th(%)	Pb(%)	204	206	207	208	Age[2]		Reference[3]
Calabogie, Ont. (Bagot Twp.)	Eu.	Tor. #63				0.0	88.88	6.66	4.50	1010 ± 70	(a)	G.S.A. Bull. 65: 1, 1954
Cassahabeg Lake, Ont.	Sy.		$\%K_2O=6.04$		$\frac{A^{40}}{K^{40}}=0.096$					1370[12] ± 90	(g)	
Cassahabeg Lake, Ont.	M.		$\%K_2O=9.51$		$\frac{A^{40}}{K^{40}}=0.062$					1020 ± 70	(g)	
Conger Twp., Ont.	Th.	Ellsworth	4.92	42.61	0.165					65	(e)	G.S.C. Econ. Geol. Ser. #11, 1932
Conger Twp., Ont.	S.	Ellsworth	11.35	1.90	0.35					210	(e)	Ibid.
Conger Twp., Ont.	U.	Ellsworth	66.10 68.81 69.18	2.94 3.12 2.83	9.76 10.76 10.83	Altered material Probably slightly altered Excellent material						Ibid.
Conger Twp., Ont.	U.	Nier	Same material as above. Uses average of previous two Ellsworth analyses. Recalculated using only best analysis			0.0057	91.81	6.79	1.40	1030 1000 1015 1010	(a) (b) (c) (d)	Phys. Rev. 60: 112, 1941
Conger Twp., Ont.	U.	Tor. #437	Identical lead to Nier			0.002	91.85	6.80	1.36	1050 ± 10	(a)	
Conger Twp., Ont.	F.	Tor. #1223A	$\%K_2O=13.0\pm0.05$ (Watson)			$\frac{A^{40}}{K^{40}}=0.055$				930 ± 70		Proc. Geol. Assoc. Can. 7: Pt. 2, 27, 1955
Conger Twp.,[7] Ont.	Mi.	Tor. #1224A	$\%K_2O=8.76\pm0.26$ (Watson)			$\frac{A^{40}}{K^{40}}=0.062$				1030 ± 80	(g)	Ibid.
Derry Twp., Que.	Th.	Ellsworth	58.45	5.94	9.17					1090	(e)	G.S.C. Econ. Geol. Ser. #11, 1932

TABLE II (cont'd)

Location and mineral[1]		Analyst	U(%)	Th(%)	Pb(%)	204	206	207	208	Age[2]		Reference[3]
Dickens Twp., Nipissing Dist., Ont.	Mo.	Ellsworth	0.27	6.43	0.27					820	(e)	Ibid.
Dill Twp., Ont.	To.	Ellsworth								420	(e)	Ibid.
Dill Twp., Ont.	F.	Tor. #1107A	9.65 $\%K_2O=12.56$ (Inman)	0.41 12.56	0.41	$\frac{A^{40}}{K^{40}}=0.052$				900 ±70	(g)	Proc. Geol. Assoc. Can. 7: Pt. 2, 27, 1955
Dill Twp., Ont.	Mi.	Tor. #1109A	$\%K_2O=9.66\pm0.18$ (Watson)			$\frac{A^{40}}{K^{40}}=0.052$				900 ±70	(g)	Ibid.
Dungannon Twp., Ont., Great Bend granite	Gr.	Tor. #1278A	$\%K_2O=5.60$ (Moddle)			$\frac{A^{40}}{K^{40}}=0.049$				850 ±70	(g)	
Frontenac Twp., Ont.	G.	Tor. #218				1.435 1.00	24.09 16.79	22.10 15.40	52.37 36.49	1350 ±185	(f)	Phys. Rev. 88: 1275, 1952
Lake Baskatong (Ciglen Claims) Que.	G.	Tor. #520				1.430 1.00	23.72 16.59	22.37 15.64	52.49 36.71	1310 ±180	(f)	Proc. Geol. Assoc. Can. 7: Pt. 2, 27, 1955
Lyndoch Twp., Renfrew Co., Ont.	Ly.	Ellsworth	0.62	4.35	0.29					1030	(e)	G.S.C. Econ. Geol. Ser. #11, 1932
Madawaska Twp., Ont.	Fe.	Tor. #134				0.011 ±0.007	90.71	6.54	2.77	960 ±50	(a)	G.S.A. Bull. 65:1, 1954
March Twp.,	U.	Ellsworth	63.79	4.32	10.77					1180	(e)	G.S.C. Econ. Geol.
Carleton Co. Ont.												Ser. #11, 1932
Mattawan Twp., Nipissing Dist., Ont.	Eu.	Ellsworth	6.02	0.85	0.98					1120	(e)	Ibid.
Monteagle Twp., Hastings Co. Ont. (MacDonald Mine)	Ut.	Ellsworth	14.63	40.72	1.23					320	(e)	Ibid.

Footnotes at end of Table.

TABLE II

DEPOSITS OF GRENVILLE PROVINCE (cont'd)

Location and mineral[1]	Analyst	U(%)	Th(%)	Pb(%)	204	206	207	208	Age[2]	Reference[3]
MacDonald Mine	Ell. Todd	15.40	—	0.22					90 (e)	Ibid.
MacDonald Mine	Ell. Todd	16.31	—	0.38					170 (e)	Ibid.
MacDonald Mine	Cy. Muench	0.529	0.080	0.043						Amer. Jour. Sci. 25: 487, 1933
MacDonald Mine	Cy. Hecht & Korisch	0.569 Same material as Muench	0.389	0.072						Mikrochemie 28: 30, 1939
MacDonald Mine	Cy. Tor. #434	Identical Pb to Hecht[8]			0.041	80.07	5.14	14.75	490 (a) 470 (b) 470 (c) 1540 (d)	Proc. Geol. Assoc. Can. 7: Pt. 2, 27, 1955
MacDonald Mine	F. Tor. #1019A	$\%K_2O = 11.97 \pm 1.2$ (Watson)			$\dfrac{A^{40}}{K^{40}} = 0.051$				880 ± 70 (g)	Ibid.
MacDonald Mine	F. Tor. #1028A	$\%K_2O = 11.8 \pm 0.24$ (Watson)			$\dfrac{A^{40}}{K^{40}} = 0.053$				910 ± 70 (g)	Ibid.
Mayo Twp., Ont.	Gr. Tor. #1279A	$\%K_2O = 5.30$ (Moddle)			$\dfrac{A^{40}}{K^{40}} = 0.044$				790 ± 80 (g)	Ibid.
Monteagle Twp., Ont. (Woodcox Mine)	Ha. Todd	13.72	0.46	0.50					260 (e)	G.S.C. Econ. Geol. Ser. #11, 1932
Woodcox Mine	Ha. Todd	9.27	0.37	0.22					170 (e)	Ibid.
Woodcox Mine	S. Ellsworth	9.32	2.94	0.44					310 (e)	Ibid.
New Calumet Mine, Ont.	G. Tor. #216				1.405 1.00	24.19 17.22	22.14 15.75	52.26 37.20	835 ±220 (f)	Proc. Geol. Assoc. Can. 7: Pt. 2, 27, 1955
North Hastings Twp., Ont.	U. Tor. #384				0.020	88.09	6.49	5.41	960 ± 30 (a)	

TABLE II (cont'd)

Location and mineral[1]		Analyst	U(%)	Th(%)	Pb(%)	204	206	207	208	Age[2]		Reference[3]
Pied des Monts, Que.	Th.	?	2.7	0.0+	0.622					1650	(e)	R.C.M.G.T. 106, 1938
Pied des Monts, Que.	U.	Muench	49.25	0.0+	6.670							J.A.C.S. 61: 2742, 1939
Pied des Monts, Que.	U.	Nier	Identical lead to Muench			0.043	91.27	6.94	1.75	918 860 867	(a) (b) (c)	Phys. Rev. 55: 153, 1939
Pied des Monts, Que.	U.	Tor. #428	Identical lead to Nier			0.044	90.91	7.14	1.80	990 ±50	(a)	
Pied des Monts, Que.	U.	Ellsworth	73.08 Not the same material as Muench	0.088	10.84							Am. Min. 19: 421, 1934
Pied des Monts, Que.	U.	Tor. #379	Identical lead to Ellsworth			0.021	92.26	6.94	0.83	1000 950 970	(a) (b) (c)	Proc. Geol. Assoc. Can. 7: Pt. 2, 27, 1955
Portland Twp., Que.	Mo.	Spencer & Muench	0.054	3.44	0.068					420	(e)	Am. Min. 20: 724, 1935
Portland Twp., Que.	Mo.	Hecht	0.094	{3.72 {3.18	{0.032 {0.05	Probably from same lot of crystals as Spencer & Muench						Ibid.
Sabine Twp., Nipissing Dist., Ont.	Eu.	Ellsworth	7.76	3.46	1.25					1010	(e)	G.S.C. Econ. Geol. Ser. #11, 1932
S. Sherbrooke Twp., Lanark Co., Ont.	Eu.	Ellsworth	7.65	2.32	0.94					800	(e)	G.S.C. Econ. Geol. Ser. #11, 1932
S. Sherbrooke Twp.	Eu.	Imperial Inst., London	8.90	1.18	0.19					150	(e)	Ibid.
Tory Hill, Ont.	Gr.	Aldrich et al.	Determinations on minerals separated from a granite							1060 1020 1050	(a) (b) (c)	Proc. Geol. Assoc. Can. 7: Pt. 2, 7, 1955

Footnotes at end of Table.

TABLE II
DEPOSITS OF GRENVILLE PROVINCE (cont'd)

Location and mineral[1]		Analyst	U(%)	Th(%)	Pb(%)	204	206	207	208	Age[2]	Reference[3]
Tory Hill, Ont. (Essonville granite)	Gr.	Tor. #1282A	%K₂O = 8.20 (Moddle)			A⁴⁰/K⁴⁰ = 0.042				760 ± 60 (g)	
Villeneuve, Que.	U.	Hillebrand	64.74	5.63	10.46	(This is probably the first chemical analysis on a Canadian mineral suitable for age determinations.)				1130 (e)	Am. Jour. Sci. 42: 390, 1891
Villeneuve, Que.	U.	Ellsworth	65.34 / 60.33 / 50.24	5.63 / 5.48 / 6.73	10.61 / 9.96 / 13.85	Centre / Middle / Outside } same crystal — Probably identical material to Hillebrand					G.S.C. Econ. Geol. Ser. #11, 1932
Villeneuve, Que.	U.	Tor. #382	Identical lead to Ellsworth (Unsatisfactory analysis)			0.094	88.03	7.33	4.57	880 (a) / 980 (b) / 950 (c) / 470 (d)	Proc. Geol. Assoc. Can. 7: Pt. 2, 27, 1955
Wilberforce, Ont.	U.	Wells	53.52	10.37	9.26					1170 (e)	J.A.C.S. 52: 4851, 1930
Wilberforce, Ont.	U.	Hecht & Reich-Rohrwig	52.68	10.17	9.70	Identical material to Wells				1240 (e)	R.C.M.G.T. 27, 1932
Wilberforce, Ont.	U.	Tor. #432	Identical material to Wells			0.002	87.89	6.63	5.49	1090 (a) / 1060 (b) / 1070 (c) / 1060 (d)	
Wilberforce, Ont.	U.	Aston	Mass spectroscopic analysis of Wells' lead			—	85.9	8.3	5.8		Nature, April 30, 1932
Wilberforce, Ont.	U.	Rose & Stranathan	Optical spectroscopic analysis of above lead			—	86.6	7.2 (207 = 207 + 204)	5.6		Phys. Rev. 50: 792, 1936
Wilberforce, Ont.	U.	Nier	Uses Wells' analysis			0.010 Cf. Rose and Stranathan, Aston and Toronto	87.98	6.59	5.42	1035 (a) / 1050 (b) / 1040 (c) / 990 (d)	Phys. Rev. 55: 150, 1939

TABLE II (cont'd)

Location and mineral[1]		Analyst	U(%)	Th(%)	Pb(%)	204	206	207	208	Age[2]		Reference[3]
Wilberforce, Ont.	U.	Tor. #16	52.99		3.87	<0.010	89.53	6.73	3.87	1050 ±20	(a)	G.S.A. Bull. 65: 1, 1954
Wilberforce, Ont.	U.	Tor. #59				0.12	82.34	7.91	9.64	1100 ±130	(a)	Ibid.
Wilberforce, Ont.	U.	Tor. #122				<0.01	89.67	6.56	3.76	1030 ±10	(a)	Ibid.
Wilberforce, Ont.	U.	Tor. #124				<0.010	89.43	6.60	3.98	1040 ±15	(a)	Ibid.
Wilberforce, Ont.	U.	Tor. #638				0.010	88.16	6.51	5.33	1020 ±20	(a)	
Wilberforce, Ont.	U.	Alter & Yuill	52.99	5.22	9.15	Outside	⎫ layers of same crystal					J.A.C.S. 59: 390, 1937
			54.47	15.25	10.06	Middle	⎬			1270	(e)	
			55.50	10.46	11.05	Inside	⎭					
			54.36	10.61	10.10	Weighted average						
Wilberforce, Ont.	U.	Hecht	56.57	10.96	9.90	Identical material to core of Alter & Yuill				1120	(e)	R.C.M.G.T. 87, 1940
Wilberforce, Ont.	U.	Todd	60.55	10.02	9.65	Very good material				1040	(e)	G.S.C. Econ. Geol. Ser. #11, 1932
Wilberforce, Ont.	U.	Ellsworth	61.43	9.32	10.16	Excellent material				1080	(e)	Ibid.
Wilberforce, Ont.	U.	Ellsworth	54.25	11.92	10.25	Altered material				1180	(e)	G.S.C. Econ. Geol. Ser #11, 1932
Wilberforce, Ont.	U.	Hecht, Kroupa & Mahr	4.63	0.58	0.91	Outside	⎫ of single crystal					R.C.M.G.T. 55, 1937
			3.61	0.45	0.89	to	⎬					
			52.85	11.91	9.04	centre	⎭					
			58.39	13.43	10.33							
			58.22	13.35	10.34							
			57.30	13.48	10.00							
Wilberforce, Ont.	U.	Alter & Kipp	37.85	8.36	9.74	Outside	⎫ layers in same crystal			1160	(e)	Science, 82: 464, 1935
			58.48	14.09	11.93	Middle	⎬				(e)	
			60.70	8.05	11.87	Centre	⎭			1340	(e)	
			47.73	9.55	11.13	Average						

Footnotes at end of Table.

TABLE II

Deposits Within or Overlying Keewatin Continental Nucleus
(<2000 m.y. old and possibly including some of Grenville age)

Location and mineral[1]	Analyst	U(%)	Th(%)	Pb(%)	204	206	207	208	Age[2]	Reference[3]
Dorion Mine, Port Arthur, Ont.	G. Tor. #853				1.364 1.00	24.96 18.29	21.70 15.91	51.98 38.10	<500 (f) Anom?	Proc. Geol. Assoc. Can. 7: Pt. 2, 27, 1955
McKellar Harbour, Ont.	G. Tor. #573				1.461 1.00	22.89 15.67	22.54 15.43	53.12 36.36	1470 ±180 (f)	Ibid.
Mattarrow Mine, Matachewan, Ont.	G. Tor. #150				1.49 1.00	23.58 15.83	22.96 15.41	51.97 34.88	1815 ±160 (f)	Ibid.
Hearst Twp., Ont.	G. Tor. #583				1.496 1.00	22.93 15.33	23.06 15.41	52.50 35.09	1840 ±160 (f)	Ibid.
New Delhi Mine, Delhi Twp., Ont.	G. Tor. #201				1.476 1.00	23.72 16.07	22.84 15.47	51.91 35.17	1660 ±170 (f)	Ibid.
Campbell-Chibougamou Mine, Que.	G. Tor. #650				1.512 1.00	22.63 14.97	23.06 15.25	52.78 34.91	1950 ±150 (f)	
Sudbury, Ont. (Fairbank Twp.)	G. Tor. #235				1.430 1.00	23.16 16.20	22.53 15.76	52.89 36.99	1205 ±180 (f)	Trans. A.G.U. 35: 301, 1954
Sudbury, Ont. Worthington Mine	G. Tor. #211				1.04 1.00	27.04 26.00	17.62 16.94	54.30 52.21	Anom.[9]	Ibid.
Sudbury, Ont. Garson Mine	G. Tor. #217[10]		Pb+ Pb(CH3)3[‡]		1.170 1.00 1.172 1.00	26.85 22.95 26.90 22.89	19.53 16.69 19.56 16.61	52.41 44.79 52.36 44.56	Anom. Anom.	Ibid. Ibid.
Sudbury, Ont. 3300' level, Frood Mine	G. Tor. #232				1.441 1.00	23.04 15.99	22.83 15.84	52.68 36.56	1355 ±185 (f)	Ibid.

TABLE II (Keewatin, cont'd)

Location and mineral[1]	Analyst	U(%)	Th(%)	Pb(%)	204	206	207	208	Age[2]	Reference[3]
Sudbury, Ont. 400' level, Frood Mine	G. Tor. #233				1.163 1.00	26.75 23.10	19.65 16.90	52.42 45.04	Anom.	*Ibid.*
Sudbury, Ont. Garson Mine	G. Tor. #234				1.171 1.00	26.93 23.00	19.45 16.61	52.44 44.78	Anom.	*Ibid.*
Sudbury, Ont. 2400' level, Falconbridge Mine	G. Tor. #305				1.150 1.00	27.25 23.70	19.46 16.92	52.15 45.35	Anom.	*Ibid.*
Sudbury, Ont. 1700' level, Falconbridge Mine	G. Tor. #306				1.140 1.00	27.59 24.20	19.32 16.95	51.96 45.58	Anom.	*Ibid.*
Sudbury, Ont. 750' level, Hardy Mine	G. Tor. #307				1.068 1.00	24.81 23.23	17.91 16.77	56.22 52.64	Anom.	*Ibid.*
Sudbury, Ont. 3300' level, Falconbridge Mine	G. Tor. #308				1.136 1.00	27.59 24.29	19.36 17.04	51.92 45.70	Anom.	*Ibid.*
Sudbury, Ont. 600' level, McKim Mine	G. Tor. #309				1.422 1.00	23.36 16.43	22.70 15.96	52.51 36.93	1095 ±200	*Ibid.*
Sudbury, Ont. 80' level, McKim Mine	G. Tor. #310				1.426 1.00	23.24 16.30	22.50 15.85	52.72 36.97	1180 ±195	*Ibid.*
Sudbury, Ont. 1000' level, McKim Mine	G. Tor. #311				1.164 1.00	26.81 23.03	19.43 16.69	52.60 45.19	Anom.	*Ibid.*

Footnotes at end of Table.

TABLE II

Deposits Within or Overlying Keewatin Continental Nucleus (cont'd)
(<2000 m.y. old and possibly including some of Grenville age) (cont'd)

Location and mineral[1]	Analyst	U(%)	Th(%)	Pb(%)	204	206	207	208	Age[2]	Reference[3]
Sudbury, Ont. 960 level, McKim Mine	G. Tor. #312				1.169 / 1.00	26.76 / 22.89	19.55 / 16.72	52.51 / 44.92	Anom.	Trans. A.G.U. 35: 301, 1954
Sudbury, Ont. Ontario Pyrites	G. Tor. #518				1.456 / 1.00	23.51 / 16.15	22.71 / 15.60	52.33 / 35.94	1455 ±180 (f)	
Theano Point, Camray Showing, Ont.	P. Tor. #17				0.20[11]	84.42	8.72	6.65	920 ±150 (a)	G.S.A. Bull. 65: 1, 1954
Theano Point, Camray Showing, Ont.	P. Tor. #18				0.18[11]	84.40	8.99	6.49	1100 ±120 (a)	Ibid.
Theano Point, Ranwick Showing	P. Tor. #78				0.104[11]	87.08	8.30	4.53	1200 ± 60 (a)	Ibid.
Theano Point, Labine-MacCarthy Showing	P. Tor. #80				0.193[11]	82.52	9.01	8.31	1100 ± 70 (a)	Ibid.
Theano Point, Labine-MacCarthy Showing	P. Tor. #137				0.052[11]	91.83	5.74	2.39	390 ±130 (a)	Ibid.
Theano Point, Labine-MacCarthy Showing	P. Tor. #138				0.096[11]	87.63	8.18	4.13	1160 ± 40 (a)	Ibid.
White River Lead Mine, Twp. 169, Ont.	G. Tor. #469				1.438 / 1.00	23.91 / 16.63	22.82 / 15.87	51.84 / 36.05	1310 ±180 (f)	Proc. Geol. Assoc. Can. 7: Pt. 2, 27, 1955

TABLE II (cont'd)

Location and mineral[1]	Analyst		U(%)	Th(%)	Pb(%)	204	206	207	208	Age[2]		Reference[3]
Cobalt, Ont. Lawson Mine	G.	Tor. #466				1.511 / 1.00	22.58 / 14.94	23.26 / 15.39	52.65 / 34.84	1975 ±150	(f)	Ibid.
Cobalt-Badger Mine, Que.	G.	Tor. #529				1.516 / 1.00	22.58 / 14.89	23.29 / 15.36	52.62 / 34.71	2010 ±150	(f)	Ibid.
Cobalt, Ont. Ingram Twp.	G.	Tor. #585				1.514 / 1.00	22.60 / 14.93	23.25 / 15.36	52.62 / 34.76	1990 ±150	(f)	Ibid.
Cobalt, Ont. Gillies Limit, Block 2	G.	Tor. #600				1.436 / 1.00	23.98 / 16.70	22.66 / 15.78	51.94 / 36.17	1270 ±190	(f)	Proc. Geol. Assoc. Can. 7: Pt. 2, 27, 1955
Cobalt, Ont. Kerr Lake Mine	G.	Tor. #601				1.501 / 1.00	22.67 / 15.10	23.27 / 15.50	52.57 / 35.02	1870 ±175	(f)	Ibid.
Cobalt, Ont. Silver Miller Mine	G.	Tor. #602				1.498 / 1.00	22.80 / 15.22	23.17 / 15.47	52.52 / 35.06	1865 ±160	(f)	Ibid.
Cobalt, Ont. Cobalt Lode Mine	G.	Tor. #603				1.351 / 1.00	24.97 / 18.48	21.53 / 15.94	52.13 / 38.59	<800	(f)	Ibid.
Cobalt, Ont. Delhi Twp.	G.	Tor. #604				1.467 / 1.00	23.43 / 15.97	22.97 / 15.66	52.13 / 35.54	1595 ±170	(f)	Ibid.
Cobalt, Ont. Kerr Lake Mine	G.	Tor. #605				1.506 / 1.00	22.64 / 15.03	23.26 / 15.44	52.58 / 34.91	1930 ±150	(f)	Ibid.

Footnotes at end of Table.

TABLE II

Deposits of Keewatin Continental Nucleus (>2000 m.y. old)

Location and mineral[1]	Analyst	U(%)	Th(%)	Pb(%)	204	206	207	208	Age[2]	Reference[3]
Barvue Mine, Que.	G. Tor. #641				1.591 / 1.00	21.48 / 13.50	23.45 / 14.74	53.47 / 33.61	2540 ±125	(f)
Chicobi Lake, Que.	G. Tor. #463				1.600 / 1.00	21.53 / 13.46	23.46 / 14.67	53.42 / 33.40	2590 ±120	(f)
Elder Mine, Que.	G. Tor. #677				1.582 / 1.00	21.62 / 13.67	23.48 / 14.84	53.31 / 33.70	2490 ±125	(f)
Golden Manitou Mine, Que.	G. Tor. #639				1.561 / 1.00	21.39 / 13.70	23.40 / 14.99	53.67 / 34.38	2330 ±140	(f)
Golden Manitou Mine, Que.	G. Tor. #661	Two different			1.597 / 1.00	21.47 / 13.44	23.45 / 14.68	53.48 / 33.49	2580 ±120	(f)
Golden Manitou Mine, Que.	G. Tor. #661	types of ore			1.602 / 1.00	21.40 / 13.35	23.45 / 14.64	53.57 / 33.44	2600 ±120	(f)
Gordonia Prop. Dalquier Twp. Que.	G. Tor. #464				1.512 / 1.00	22.66 / 14.99	23.19 / 15.34	52.65 / 34.82	1970 ±150	(f)
Horne Mine, Noranda, Que.	Mag. Hurley	dœs/hr./cm² = 0.26			He (10⁻⁵cc/gm) =		11.4		2000	Science, *110*: No. 2845, 49, 1949
Huron Claim, Winnipeg River, Man.	U. Hecht	38.41 / 38.49	9, 11 / ?	13.67 / 13.64					2370	(e) R.C.M.G.T. 18, 1933
Huron Claim, Winnipeg River, Man.	Mo. Kroupa & Hecht	0.12	12.67	1.21					2080	(e) Zeit. Anal. Chem. *106*: 82, 1936
Huron Claim	U. Ellsworth & DeLury	53.50 / 55.01	12.46 / 12.25	15.44 / 15.50	Better material — Both analyses possibly low in lead — cf. reference					Am. Min. *16*: 569, 1931
Huron Claim	U. Nier				0.018	81.50	13.18	5.31	2490 (a) / 1370 (b) / 1980 (c) / 1270 (d)	Phys. Rev. *55*: 150, 1939

Uses average of Ellsworth & DeLury.
Recalculated using better analysis.
Age (a) appeared in error in original paper.

TABLE II (cont'd)

Location			Description / Data	Values				$\frac{A^{40}}{K^{40}}$ / He	Age	Note	Reference
Huron Claim	Mo.	Muench		0.281	15.63	1.524			2600 / 3200 / 2850 / 1860	(a) (b) (c) (d)	G.S.A. Bull. *61*: 129, 1950
Huron Claim	Mo.	Nier	Identical lead to Muench	0.0097	10.2	1.86	87.93				Phys. Rev. *60*: 112, 1941
Huron Claim	U.	Tor. #429	Identical lead to Nier	0.032	80.31	13.50	6.10		2535 ±100	(a)	Proc. Geol. Assoc. Can. *7*: Pt. 2, 27, 1955
Huron Claim	U.	Tor. #504	Very fresh new material	0.045	79.83	14.02	6.11		2580 ±100	(a)	*Ibid.*
Huron Claim	Al.	Tor. #1240A	%K_2O = 1.60					$\frac{A^{40}}{K^{40}} = 0.211$	2330 ±150	(g)	
Huron Claim	M.	Tor. #1241A	%K_2O = 9.15					$\frac{A^{40}}{K^{40}} = 0.234$	2470 ±150	(g)	
Falcon Island Lake of the Woods, Man.	Le.	Ahrens	Spectroscopic determination of Sr/Rb Approximate results only. Error > 10%						2200	(h)	G.S.A. Bull. *60*: 217, 1949
Lacorne Twp., Que.	Mi.	Tor. #1116A	%K_2O = 10.0 (Cumming)					$\frac{A^{40}}{K^{40}} = 0.242$	2510 ±150	(g)	
Lacorne Twp.	Mi.	Tor. #1120A	%K_2O = 10.6 (Cumming)					$\frac{A^{40}}{K^{40}} = 0.238$	2490 ±150	(g)	
Larder Lake Dist., Ont.	Mag.	Hurley	α's/hr./cm^2 = 0.36					$He(10^{-5}cc/gm) = 16.9$	2100		
Larder Lake Dist., Ont.	Mag.	Hurley	α's/hr./cm^2 = 0.15					$He(10^{-5}cc/gm) = 8.4$	2400		
Silver Leaf Mine, Man.	Le.	Ahrens	Spectroscopic determination of Sr/Rb Approximate results only. Error > 10%						2100	(h)	G.S.A. Bull. *60*: 217, 1949
Silver Leaf Mine, Man.	Le.	Ahrens	As above						2350	(h)	*Ibid.*

Footnotes at end of Table.

TABLE II

DEPOSITS OF KEEWATIN CONTINENTAL NUCLEUS (>2000 m.y. old) (cont'd)

Location and mineral[1]	Analyst	U(%)	Th(%)	Pb(%)	204	206	207	208	Age[2]	Reference[3]
Along the Winnipeg River, Man.	Le. Ahrens	As above							2300 (h)	Ibid.
Winnipeg River, Man.	Le. Ahrens & Gorfinkle	Refined spectroscopic technique							2400 (h)	Nature, 166: 149, 1950
Silver Leaf Mine, Man.	Le. Tor. #1080A	%K₂O = 10.1 (Sturm)			$\frac{A^{40}}{K^{40}} = 0.260$				2600 ±150 (g)	Proc. Geol. Assoc. Can. 7: Pt. 2, 27, 1955
Silver Leaf Mine, Man.	Grei. Tor. #1025A	%K₂O = 9.14			$\frac{A^{40}}{K^{40}} = 0.204$				2280 ±150 (g)	
Geneva Lake Mine, Ont.	G. Tor. #515				1.545 1.00	21.99 14.23	23.42 15.16	53.05 34.34	2230 ±140	Proc. Geol. Assoc. Can. 7: Pt. 2, 27, 1955
Geneva Lake Mine, Ont.	G. Tor. #516				1.550 1.00	21.87 14.11	23.41 15.10	53.18 34.31	2305 ±135 (f)	Ibid.
Hollinger Mine, 400' level, Ont.	G. Tor. #525				1.589 1.00	21.43 13.49	23.44 14.75	53.56 33.71	2515 ±125 (f)	Ibid.
Lakeshore Mines, Kirkland Lake, Ont.	G. Tor. #519				1.553 1.00	22.32 14.37	23.33 15.02	52.79 33.99	2280 ±140 (f)	Ibid.
Manitouwadge, Ont.	G. Tor. #659				1.598 1.00	21.44 13.42	23.41 14.65	53.54 33.51	2570 ±120 (f)	
Montbeillard Twp., Timiskaming Co., Que.	G. Tor. #524				1.570 1.00	21.85 13.91	23.55 15.00	53.07 33.80	2400 ±130 (f)	
McElroy Twp., in Timiskaming sediments	G. Tor. #584				1.518 1.00	22.80 15.02	23.17 15.26	52.51 34.59	2010 ±150 (f)	Proc. Geol. Assoc. Can. 7: Pt. 2, 27, 1955
Noranda, Que. New Norzone Property	G. Tor. #582				1.566 1.00	21.86 13.96	23.42 14.96	53.16 33.95	2365 ±135 (f)	Proc. Geol. Assoc. Can. 7: Pt. 2, 27, 1955

TABLE II (cont'd)

Location and mineral[1]	Analyst	U(%)	Th(%)	Pb(%)	204	206	207	208	Age[2]	Reference[3]
Quemont Mine, 12th level, Noranda, Que.	G. Tor. #367				1.218 / 1.00	25.90 / 21.26	19.69 / 16.17	53.20 / 43.68	Anom.	Ibid.
Quemont Mine	G. Tor. #368				1.206 / 1.00	25.08 / 20.80	19.52 / 16.19	54.18 / 44.93	Anom.	Ibid.
Sioux Lookout, Ont.[10]	G. Tor. #149		Pb+ $Pb(CH_3)_3^+$		1.570 / 1.00 / 1.569 / 1.00	22.07 / 14.06 / 21.97 / 14.00	23.33 / 14.86 / 23.41 / 14.92	53.02 / 33.77 / 53.04 / 33.80	2405 ±130 (f)	Science, 118: #3069, 486, 1953
Steeprock Lake, Ont.	G. Tor. #404				1.574 / 1.00	21.94 / 13.94	23.35 / 14.83	53.14 / 33.75	2415 ±130 (f)	Ibid.
Timmins, Ont.	G. Tor. #472				1.579 / 1.00	21.70 / 13.75	23.48 / 14.87	53.24 / 33.72	2455 ±125 (f)	Proc. Geol. Assoc. Can. 7: Pt. 2, 27, 1955
Upper Canada Mine, Ont.	G. Tor. #146				1.44 / 1.00	28.22 / 19.60	21.92 / 15.22	48.43 / 33.63	Anom.	Ibid.

[1]The mineral types are denoted by the following symbols:

Al. = albite	G. = galena	Ha. = hatchettolite	Mag. = magnetite	P. = pitchblende	Th. = thucolite	Ut. = uranothorite
Cy. = cyrtolite	F. = feldspar	Le. = lepidolite	Mi. = mica	S. = samarskite	To. = toddite	
Ell. = ellsworthite	Fe. = fergusonite	Ly. = lyndocite	Mo. = monazite	Sy. = syenite.	U. = uraninite	

Eu. = euxinite   Gr. = granite
Fe. = fergusonite   Grei. = greissen

[2](a) —age calculated from 207/206 ratio (years×10⁹).
(b) —age calculated from 206/238 ratio.
(c) —age calculated from 207/235 ratio.
(d) —age calculated from 208/232 ratio.
(e) —chemical age.
(f) —age calculated from lead isotope abundances.
(g) —age calculattd from A⁴⁰/K⁴⁰ ratio.
(h) —age calculaeed from Rb/Sr ratio.

Ages have been recalculated using revised constants and dating curves (see Cumming et al., 1955).
[3]Where no reference is indicated for analyses by the Toronto Laboratory, the results have not been previously published.
[4]Toronto sample number.
[5]K₂O determined by analyst given in brackets.
[6]A⁴⁰/K⁴⁰ = 0.0561 as recalculated by Wasserburg (personal communication); age is therefore 940 million years.
[7]The feldspar and mica are from the same pegmatite.
[8]This material is probably the same as that used in the more reliable analysis by Muench. Ages have been calculated using the data of Muench, but notice the high 208/232 age and the large discrepancy between Muench and Hecht for Th. The analyses of Hecht and his co-workers are michrochemical analyses and in many cases may not be reliable.
[9]Anomalous lead. See Russell et al., Trans. A.G.U., 35: 301, 1954 for methods of dating this type of lead.
[10]Two analyses were run on this sample using the Pb+ and Pb(CH₃)⁺³ spectra. Unless otherwise indicated, all Toronto results are based on analyses of the Pb(CH₃)⁺³ spectrum which is believed to be more reliable.
[11]Corrected for common lead using Anacon Mine analyses.
[12]All potassium-argon ages reported here have been calculated using a branching ratio of 0.089, based on experimental evidence of potash minerals. It is possible that for some minerals and rocks, a ratio of 0.12 should be used which would decrease the age of the syenite to 1130±80 million years.

APROPOS THE GRENVILLE

A. E. Engel

DATA AND IDEAS apropos things Grenville have been produced for a century by many people of varied backgrounds and interest. The record is long and complex. Small wonder that in reading it serious discrepancies appear and major difficulties in separating inference from fact are encountered. Yet an attempt to detach the facts and to assess the ideas may focus attention and work upon important problems of the future. One means of attempting an analysis of pertinent Grenville problems is to focus attention upon the uses and implications of terms such as Grenville series, Grenville subprovince, Grenville front, and Grenville orogenic belt.

CORRELATIONS AND CONCEPTIONS OF THE GRENVILLE SERIES

Sections at and Southwest of the Type Locality

The term "Grenville series" appears to have been introduced most inauspiciously by Logan in the supplement to his Annual Report of 1863 (p. 839). In preceding pages of this and earlier reports (1863, p. 43; 1853–56, pp. 7–38) Logan employed the term "Grenville band" for the marble unit in the vicinity of Grenville village. There is no doubt, however, that Logan used the term Grenville series for the complex of marbles (and associated quartzite, amphibolites and garnetiferous gneiss) along and immediately northeast of the Ottawa River, especially between the tributaries Rouge and Nord (1853–56, see especially the map "Showing the distribution of crystalline limestones of the Laurentian series"). Previously Logan had referred to these "altered sedimentary deposits" as the "Laurentian System," "Laurentian Series," and "Laurentian Formation" (pp. 7–38, 1856). They are depicted on his map of 1853 as the "Laurentian series" (1856).

Logan (1863, pp. 838–839) was quite objective about the relations of the various marble units. He carefully indicated the difficulties of correlation and the alternative age possibilities suggested by the field relations. He also speculated upon the existence of a major unconformity within the "Grenville series" not far from the type locality (p. 839, 1863).

If rigorous stratigraphic methods and concepts are applied to the Grenville series, its relations and distribution would remain essentially as outlined in Logan's speculations. For despite the subsequent work by hundreds of succeeding geologists in the vicinity of the type locality and away from it,

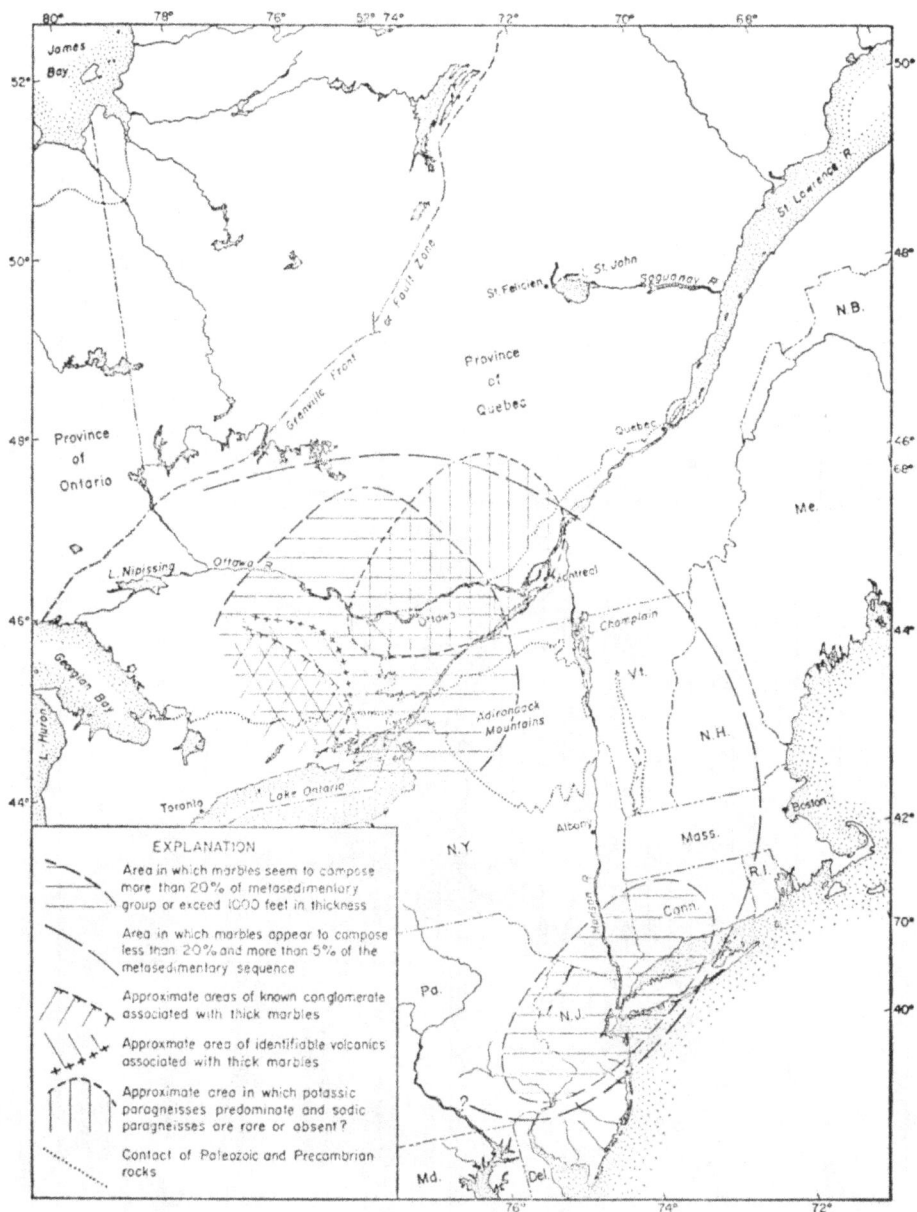

FIGURE 1. Sketch map showing the known distribution of major Precambrian marble units, with associated conglomerates, identifiable volcanics, and dominantly potassic paragneisses in the Grenville subprovince and possible outliers.

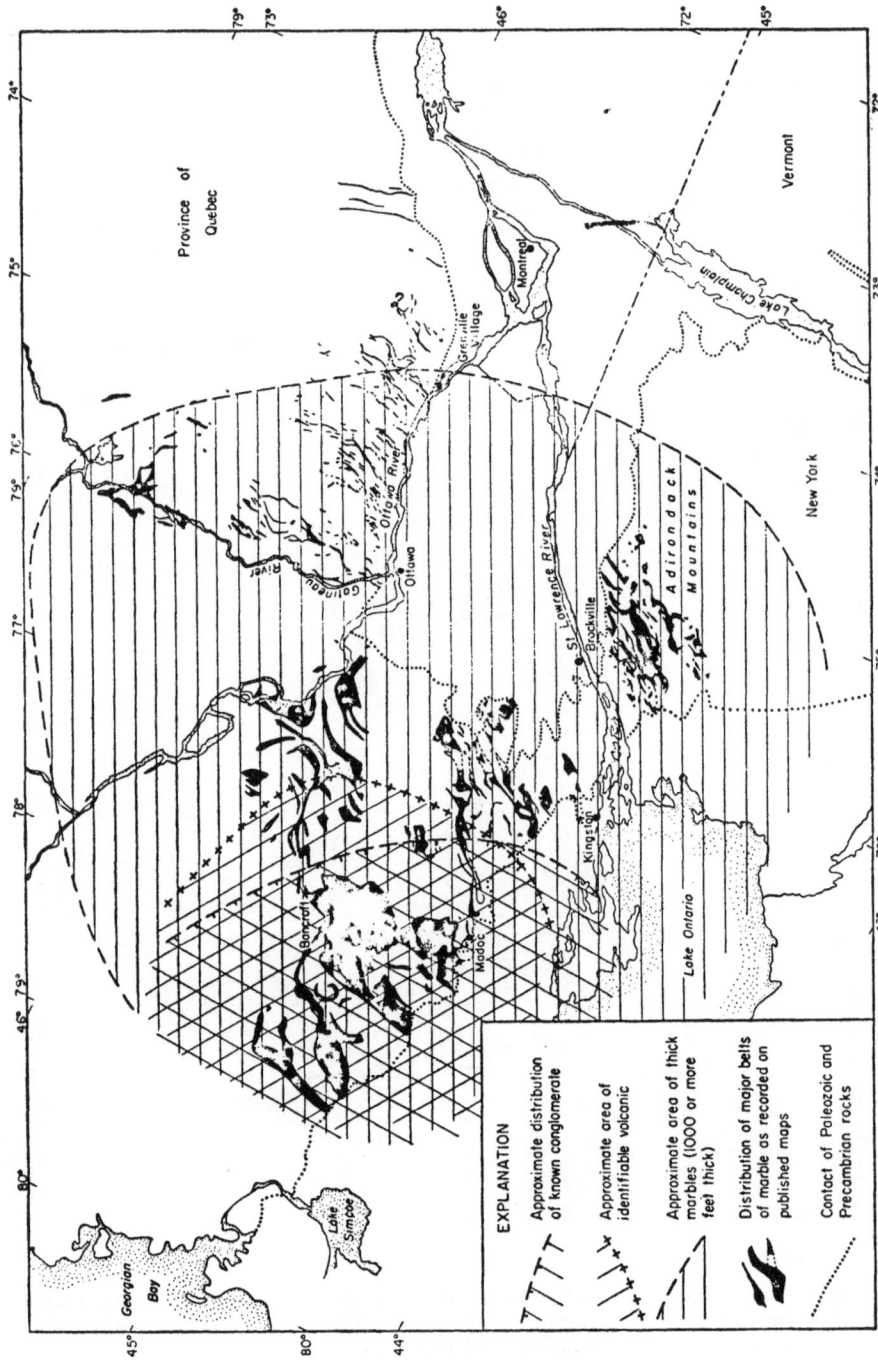

EXPLANATION

⊥⊥⊥⊥⊥ Approximate distribution
         of known conglomerate

✗✗✗✗ Approximate area of
         identifiable volcanic

╱╱╱╱ Approximate area of thick
         marbles (1000 or more
         feet thick)

▰▰▰ Distribution of major belts
         of marble as recorded on
         published maps

∙∙∙∙∙ Contact of Paleozoic and
         Precambrian rocks

FIGURE 2. Known distribution of thick marbles, and associated conglomerates and identifiable volcanics in the Grenville subprovince.

before either many years of painstaking study[2] or some miraculous means of correlating highly metamorphosed rocks. Conceivably a myriad of stratigraphic and structural booby traps, including important angular unconformities, may lie within the Grenville metasedimentary complex in the area of possibly contiguous marbles in Figure 1.

Associated Igneous Rocks.

The igneous complexes associated with the "Grenville metasedimentary rocks" are, of course, the anorthosite-gabbro-shonkonitic or charnokitic syenite, quartz-syenite and granite suites, more or less apparent or inferred at widely separated areas (Buddington, 1939, 1948; Logan, 1865; Osborne, 1936; Wilson, 1925; etc.).

To many workers the association of rocks of this igneous suite with the marble-quartzite-amphibolite-garnetiferous paragneiss sequence is most compelling. Logan grouped anorthosites, syenites, and granites with his "Grenville Series" as the "Laurentian System" (1865). Morley Wilson's definition of the Grenville subprovince also refers to a "complex" in which he includes the Grenville Series and the "related series of igneous intrusives —peridotite, diorite, shonkonite, etc." (1925, p. 389).

It is at least implicit in the writings of many workers and specifically suggested by J. Tuzo Wilson (1949, pp. 2–6), that these igneous rocks, like the metasedimentary units, are manifestations of the crustal mechanics distinctive of the evolution of the Grenville subprovince. If a close (genetic) association of, say, marble-quartzite-amphibolite-garnetiferous paragneiss with anorthosite-syenite and granite is assumed, spot correlations become more tempting. Certainly the association of these seven rock types at any point not too many hundreds of miles from the valleys of the Ottawa and Rouge rivers would tease a tentative correlation of some kind out of the most Grenoble of us all.

This is why geologists who have studied both the Grenville Series in the marble rich area of Figure 1, and the rocks in the Precambrian highlands of New Jersey, commonly think of the New Jersey area as another outlier of the Grenville subprovince. This possible relationship is referred to again in a succeeding part of this paper.

Sections Northeast of the Type Locality

But what about map areas within the subprovince per se, northeast of the marble-rich Grenville Series? Many of these areas of Quebec are separated from the type locality by unmapped or little understood geology.

[2]J. S. Brown and the author have in press a "Revision of the stratigraphy and structure of the Edwards-Balmat District," in which a marble of the "Grenville series" some 2000 feet thick is differentiated into sixteen remarkably persistent stratigraphic units. The number of units could be doubled by further subdivisions if this were desirable. These units are distinctive and mappable in most places where extensive rock exposures and subsurface information are available. But this stratigraphic reconstruction required a map of the district at 800 feet to the inch, maps of miles of underground workings at much larger scales, as well as the detailed data from over 100,000 feet of drill core. The time involved to reconstruct the detailed stratigraphy and structure of a few square miles of marble near Balmat, N.Y., exceeds 10 man years.

There is not the beguiling or reassuring "band" of marble more or less continuous with that near Grenville village. Indeed, in many of these map areas of Quebec, marble either is not present or occurs only in isolated, thin lenses (Fig. 2). There are, of course, other Grenoble rocks present: one or more types such as garnetiferous paragneisses, amphibolites, quartzites as well as possibly anorthosite, syenite, and the ubiquitous granite.

TABLE I

PROVISIONAL TABULATION OF MAJOR PRECAMBRIAN METASEDIMENTARY AND VOLCANIC
ROCK GROUPS IN THE "GRENVILLE SUBPROVINCE"

These three groups appear to represent distinctive metamorphic complexes as well as contrasting sedimentary rock types. Data largely from published maps and reports of the Geological Survey of Canada and the Ontario and Quebec Bureau of Mines.

Region and type	Haliburton-Kingston region (Madoc group or Kalador group?)		Ottawa River region (type locality Grenville series; Grenville group?)		St. Félicien-Saguenay-St. Morin River region (Saguenay or Moisie group? St. Félicien group?)		Region and type
Metamorphic rock	Mean	Range	Mean	Range	Mean	Range	Inferred parent rock
Marble	54	30-80	35	20-60	1	0-10	Limestone and dolomite
Quartzite	6	2-20	12	2-25	25^x	10-45^x	Cleaner, well sorted ss.
Feldspathic gneiss (Paragneiss)	24$^+$	10-40	45$^°$	20-50	54$^°$	30-80	Impure ss, or shale
Amphibolite†	8	2-30	8	2-20	20	5-35	Sills, lavas, tuff, dolomite?
Conglomerate	2	0-10	—	—	Rare	0-1	Conglomerate
Volcanic*	6	0-20	ne	ne	ne	ne	Pillow lava and tuff

x = includes impure types
+ = includes diverse types?
● = largely shale or argillaceous sandstone?
° = largely graywacke and tuff?

† = origin problematical
* = origin known, or readily inferred
ne = no definite examples

But in many, if not most, of these map areas certain of the Grenvillianous ("Keewatin" or "Timiskaming") rock types also appear or dominate, especially metavolcanics, and meta- or tuffaceous? greywacke-like rocks. Each worker in his area, more or less isolated from any time or rigorous lithologic correlation, is faced with balancing what to him are Grenoble versus Grenvillianous aspects of the geology. It is natural that the extent of the Grenville series and the size of the subprovince may depend upon the willingness of the worker to be thoughtful, imaginative, extremist, conservative, or outrageous.

It is a most commendable fact that most of the workers in these regions have maintained a hold on objectivity, even though they were in Precambrian rocks southeast of the "Grenville front." This part discontinuity, part phenomenon of the draughtsman's table appears to loom less high in parts of Quebec (Deland, 1953; Grenier, 1953; Gilbert, 1952; Osborne[3]) than in Ontario (Johnston, 1954). To be sure, the Grenville front exists in part because of nature's collusion with man's credulity. The rocks, ore deposits, general degree of metamorphism, and the tectonic trends are different—in some degree—to the northwest. But the interesting and disconcerting transitions in rock types along and southeast of the "Grenville front" may cause some of us to revise our conception or definitions of the Grenville series and the Grenville subprovince.

For example, what are the major components of the "Grenville series" in the eastern one-half to one-third of the Grenville subprovince? The major metasedimentary rocks, as indicated in the published reports and map sheets, seem to be gneisses and amphibolites (Table I). Both of these rock types deserve thoughtful analysis. In most localities they seem to be thoroughly metamorphosed, and referable to the amphibolite or granulite rank of metamorphic facies. The paragneisses commonly are reported to show relict bedding. but most other textural features suggestive of their origin are either obliterated or unreported. Consequently the principal clue we have regarding their origin in addition to their gross field associations and thickness, is their chemical composition. Although few chemical or modal analyses of these paragneisses exist in the literature, the meagre data suggest a dominance of interesting soda-rich types not unlike the major Adirondack paragneiss (Engel and Engel, 1953b). The dominant mineral constituents are quartz, biotite, and plagioclase feldspar (commonly with garnet). Two published analyses of these gneisses, from Gillies (1952), are listed in Table II (analyses 2 and 3) alongside analyses of "Keewatin" and "Timaskaming" type and "Archean" schists and greywackes. The similarities between the so-called "Grenville type" gneisses in Quebec (analyses 2, 3), the Adirondacks (analysis 1), and the Keewatin and Timaskaming greywackes (analyses 4, 6, 8) are striking.

The excess of Na_2O over K_2O in these gneisses as contrasted with the low ratio of $Na_2O:K_2O$ in most typical shales and sandy shales is of particular interest. Although the high ratio of $Na_2O:K_2O$ in many of the so-called "Grenville-type" gneisses is by no means proof of their origin, it certainly suggests a greywacke or tuffaceous greywacke as the parent sediment rather than a shale or sandy shale formed as a product of residual weathering. The reasons for this conclusion have been noted earlier (Engel and Engel, 1953b). The difference in $Na_2O:K_2O$ involved are apparent from inspection of Figures 3 and 4. In Figure 3, the $Na_2O:K_2O$ ratio of 35 Pre-

[3]Written communication. These comments were prepared before the contents of Dr. Osborne's contribution to this symposium were known to the author. His data offer further substantial evidence of what has been implicit in the literature for a long time.

TABLE II

ANALYSES OF SODIC AND OF POTASSIC PARAGNEISSES FROM METASEDIMENTARY SEQUENCES IN THE GRENVILLE SUBPROVINCE, LISTED WITH AN "AVERAGE" SANDY SHALE AND SOME TYPICAL GREYWACKES

	1	2	3	4	5	6	7	8	9
SiO_2	70.90	66.24	63.50	65.00	67.42	61.52	69.69	60.51	68.1
TiO_2	.32	.61	.66	.58	15.74	0.62	0.40	0.87	0.7
Al_2O_3	12.17	15.42	17.32	16.10		13.42	13.53	13.56	15.4
Fe_2O_3	1.31	.76	1.06	1.08	.35	1.72	0.74	0.76	1.0?
FeO	4.12	5.40	5.40	3.78	3.46	4.45	3.10	7.63	3.4
MnO	.04	.09	.09	.08		—	.01	0.16	0.2
MgO	2.32	3.21	3.68	2.71	1.86	3.39	2.00	3.39	1.8
CaO	1.55	3.15	2.64	2.49	3.36	3.56	1.95	2.14	2.3
Na_2O	3.74	3.19	3.21	4.54	2.94	3.73	4.21	2.50	2.6
K_2O	2.87	1.43	1.85	2.06	1.73	2.17	1.71	1.69	2.2
H_2O^+	0.21	.43	.21	.13	} 3.55	2.33	2.08	3.38	} 2.1
H_2O^-	.05	.08	.05	.16		0.06	0.26	0.15	
P_2O_5	—	.08	.05	.11		—	0.10	0.27	0.08
CO_2	—	—	—	.06	.31	3.04	0.23	1.01	—
S	—	.26	.22	.15				.42	0.05
Total	99.60				99.72				

1. Composite sample of 24 least altered layers of quartz-biotite-oligoclase gneiss (major paragneiss, N.W. Adirondacks), Ledoux and Co., analysts.
2. Garnet-biotite-plagioclase-quartz hypersthene gneiss, south of the Kewatin-Timiskaming "boundary" in the Canimiti River Area (Gillies, 1952, 0. 19). Analysts H. Boileau and J. Gagnon.
3. Garnet-biotite-plagioclase-quartz gneiss as above (analysis II) Canimiti River Area, Pontiac County (From Gillies, 1952, p. 21). Analysts H. Boileau and J. Gagnon.
4. Garnet-biotite-plagiocalse-quartz gneiss ("greywacke") along the northern margin of the "Grenville subprovince." Courtesy, F. Fitz Osborne, analyst not cited.
5. Quartz-biotite-feldspar gneiss associated with marble about 4 miles northeast of Madoc, Ontario (from Miller and Knight, 1914, p. 80). Analyst not cited.
6. Precambrian greywackes (average of three). After Todd, 1928, p. 20.
7. Franciscan (Jurassic) greywackes (average of three). After Taliaferro, 1943, p. 136.
8. Precambrian greywacke, Manitou Lake, Ontario, B. Bruun, Analyst. After Pettijohn, 1949, p. 250, Table 64.
9. Average of thirty greywackes compiled by Tyrrell, 1933, p. 26, cited by Pettijohn, 1949, p. 250, Table 64.

cambrian slates (shales and siltstones) are plotted in one histogram below a similar plotting of 30 greywackes and tuffaceous greywackes. The data on Precambrian slates are from the recent summary by Nanz (1953). The data on greywackes represent most of the analyses available in the literature (16 sources). Figure 4 is an analogous histogram of some "Grenville-type" paragneisses.

The fact that in the so-called "Grenville-type paragneiss" the metamorphic mineral assemblage may differ from that of some "Keewatin" or "Timiskaming" rocks to the northwest, does not preclude their identical chemical composition or origin in similar sedimentary environments. Other examples of this apparent similarity in composition between certain Grenville type paragneisses and the rocks along or northwest of the "Grenville front," have been discussed elsewhere (Engel and Engel, 1953b, pp. 1091–1095). We may ask which of these or other paragneisses is the "Grenville type"? That at the type locality appears to be potassic. Southwest of

TABLE II (cont'd)

ANALYSES OF SODIC AND OF POTASSIC PARAGNEISSES FROM METASEDIMENTARY SEQUENCES IN THE GRENVILLE SUBPROVINCE, LISTED WITH AN "AVERAGE" SANDY SHALE AND SOME TYPICAL GREYWACKES

	10	11	12	13	14	15	16	17	18
SiO_2	65.5	61.96	57.66	74.70	58.68	66.33	49.61	55.18	58.68
TiO_2	.5	1.66	—	—	1.39	1.52	2.00	1.72	1.39
Al_2O_3	12.00	19.73	22.83	8.88	16.17	17.17	22.00	20.70	16.17
Fe_2O_3	3.10	{FeS–4.33		9.64	1.66	3.93	1.93	4.50	1.66
FeO	1.7	{4.60			5.69	6.55	9.55	5.65	5.69
MnO		trace		0.5		.09	.05	.05	
MgO	2.0	1.81	3.56	1.87	3.71	3.35	6.33	3.49	3.71
CaO	4.5	.35	1.16	1.07	.30	.90	.36	.34	.30
NA_2O	1.0	.79	.60	.42	.83	.73	1.40	1.81	.83
K_2O	2.6	2.50	5.72	.95	8.68	4.5	3.88	4.75	8.68
H_2O^+	2.9		{1.50	{1.05	{1.65	{1.00	2.20	1.80	1.65
H_2O^-		{1.82					—	——	
P_2O_5	0.1				.31	.04	.06	.10	0.31
CO_2	3.4	—	—	—	.36				.36
S	.3	—	—	—			—	—	
Total	99.6	99.55	100.77	99.08	99.43	100.18	99.37	100.09	

10. One part average sandstone plus two parts average shale (Pettijohn, 1949, p. 271).
11. Gneiss from about one mile west of St. Jean de Matha. Contains garnet, sillimanite, quartz, orthoclase with accessory graphite and pyrite. Interbedded with quartzite (from Adams, 1895, p. 58).
12. Gneiss from west shore Trembling Lake. Contains quartz, orthoclase, biotite, sillimanite. Stratigraphically near marble (from Adams, 1895, p. 58J). Analyst not cited.
13. Gneiss from Darwins Falls, near village of Rawdon, Quebec. Garnetiferous, quartzitic and interlayered with quartzite (from Adams, 1895, p. 58J). Analyst not cited.
14. Quartz-biotite-K-feldspar gneiss (with garnet and sillimanite) from south shore of the Ottawa River, Prescott County, Ontario, as cited by Morley Wilson, 1925. Analyst M. F. Conner.
15, 16, 17. Gneisses from Buckingham Township, Papineau County Quebec, as cited by Morley Wilson (1925). Analyst M. F. Conner.
18. Quartz-biotite-K-feldspar gneiss with accessory zircon, tourmaline, muscovite, apatite and magnetite. Right bank of the Ottawa River, south of Montebello. From Osann, 1901, p. 5–0. Analyst, a Dr. Dittrich.

the type locality the two general types exist. Northeast of it sodic paragneisses seem to predominate. To call a gneiss "Grenville type" without clearly designating its nature is to invite misunderstanding and chaos.

Interestingly enough the recent work of F. Fitz Osborne[4] and associates (Grenier, 1953; Deland, 1953; Gilbert, 1952) verifies what seemed implicit in earlier reports: that the (sodic) quartz, biotite plagioclase gneisses often designated as "Grenville type" northeast of the type locality commonly occur in sequences in which marble and quartzite are thin or absent. Any such mutually exclusive relationship between sodic gneisses on one hand, and marble and quartzite on the other, would be consistent with a greywacke or tuffaceous origin for the gneisses.[5]

The other major "Grenville type" in many parts of the northern and

[4] Written communication.
[5] Curiously enough there is no such comforting separation of sodic paragneiss from marble in southernmost Ontario or the Adirondacks. The two are interlayered and seemingly part of a closely related sedimentary group.

eastern parts of the "Grenville subprovince" is amphibolite. Most are evolved from parent rocks of unknown origin (de La Rue, 1948, pp. 18–19; Faessler, 1948, p. 9; Satterly, 1942, p. 7; Gilbert, 1952, pp. 3–4; Grenier, 1953, pp. 3–4; Grieg, 1945, pp. 9–10; Retty, 1944, p. 12). Unlike the paragneisses, neither the bulk composition of the amphibolites nor their present metamorphic mineral constituents, suggest a particular origin.

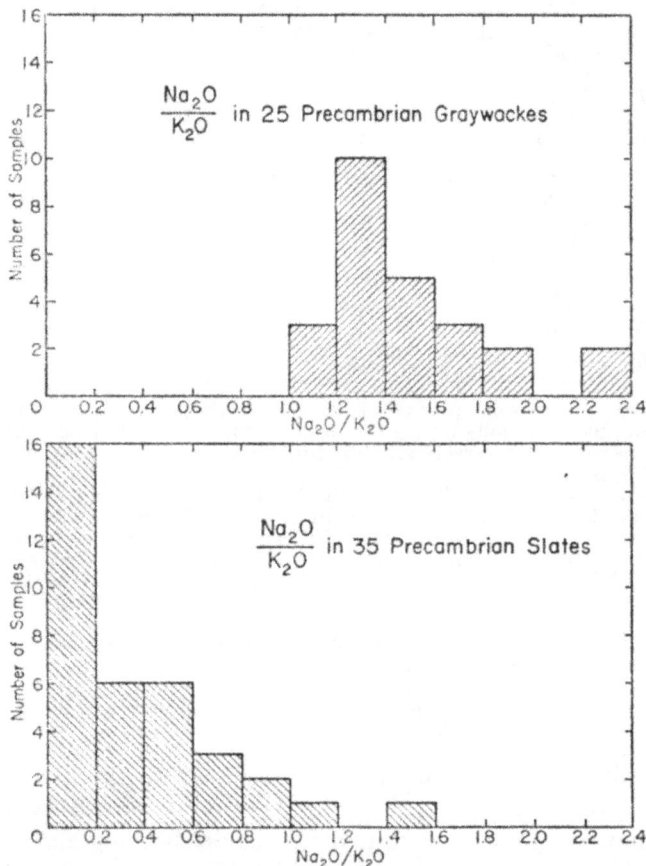

FIGURE 3. Histograms representing the frequency distribution of the ratio $Na_2O:K_2O$ in 34 Precambrian slates and in 25 greywackes. Data on slates from Nanz, 1953. Data on greywackes from 16 sources. The slight overlap in the ratio of $Na_2O:K_2O$ in the two rock types disappears completely if the selection of slates is confined to those found interlayered with appreciable limestone or dolomite and clean quartzite.

For example, in Table III are grouped nine analyses of amphibolites from the marble-rich "Grenville series" of Ontario and the Adirondacks, along with the analysis of Daly's average gabbro. It will be noted that three of the analyses are of feather amphibolite (from the Bancroft area) which can be proven from field relations to have formed by metasomatic replacement of marble. Analyses SK 2–7 and SK 4–6 are examples of Adirondack amphibolites which also definitely have formed by metasomatic alteration

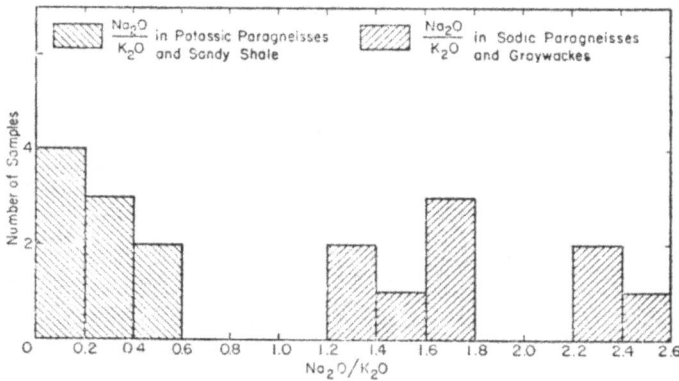

FIGURE 4. Histogram representing the distribution of the ratio $Na_2O:K_2O$ in paragneisses from the Grenville subprovince. Chemical analyses of these gneisses and their locations are given in Table II.

of dolomite marble. In contrast, the analyses Am 166 and Am 167 are from the margins of large deformed gabbro sheets in the Adirondacks. The overlapping compositions of these and various other amphibolites prevent any diagnosis of their origin if relics of the parent rock and transitional types were destroyed by metamorphism. This fact is suggested by Figure 5 in which the major constituent oxides of "Grenville" amphibolites of known and unknown origin are plotted on a triangular diagram.

Although most geologists now recognize that chemical and mineralogical composition of most amphibolites in the "Grenville subprovince" is no clue to their origin, amphibolites have remained Grenoble rocks. Actually their presence with quartz-biotite plagioclase gneiss (garnetiferous or not) could as well reflect intermediate or higher rank metamorphism of a typical "Keewatin" or "Archean" sequence as anything "Grenville." Indeed this seems to be what has happened[6] in several areas recently described.

As implied above, one of the working definitions of the "Grenville series" is any Precambrian rock in Canada southeast of the "Grenville front." It would seem that geographic position alone tends to colour our basic assumptions and our interpretations.

Recognizably, those who accept the association of both igneous-looking and metasedimentary rocks as typically "Grenville series" or "Grenville subprovince" could argue that it is the presence of anorthosites or syenites or both, with the amphibolite-paragneiss sequence, which defines much of the northeastern one-third of the Grenville subprovince. Yet not one of these rocks, nor the complete association of anorthosite-syenite, amphibolite-paragneiss (or marble), is temporarily or spatially confined to the "Grenville subprovince" or to any shield area. Are there Grenville-type anorthosites or syenites which may be distinguished from, respectively,

[6]At this symposium F. Fitz Osborne concludes that many of the sodic paragneisses and amphibolites in the northeastern one-half of the "Grenville subprovince" are indeed "Timiskaming" types.

TABLE III

TYPICAL AMPHIBOLITES DERIVED FROM BOTH IGNEOUS PARENT ROCKS AND BY
REPLACEMENT OF MARBLE OF THE "GRENVILLE SERIES"
These are listed for comparison alongside Daly's average gabbro (olivere free)

					SPECIMEN					
Oxide	SK2-7	SK4-6	FA-4	39-L	Am2	Am3	Am166	Am167	AAm1	D.G.
SiO_2	45.01	47.25	47.33	43.84	50.00	50.83	44.20	51.66	49.88	49.50
Al_2O_3	11.98	13.38	17.61	16.63	18.84	18.64	14.22	16.85	14.41	18.00
Fe_2O_3	1.88	2.10	1.57	3.50	2.57	2.84	2.27	2.48	3.32	2.80
FeO	10.65	12.10	9.95	10.24	5.51	5.97	11.51	6.48	8.60	5.80
MgO	2.61	7.52	7.39	8.86	4.63	4.90	9.35	5.94	7.70	6.62
CaO	17.60	10.96	7.15	10.04	10.65	7.50	7.68	8.70	8.74	10.64
Na_2O	1.96	2.39	4.21	1.99	4.46	4.22	2.68	2.91	1.68	2.82
K_2O	1.10	.61	.17	1.32	1.18	1.83	2.96	1.64	1.79	.98
H_2O+	.63	.92	1.10	1.70	1.00	1.40	1.17	1.12	1.69	
H_2O-	.04	.07	.05				.07	.05	.03	
CO_2	3.84	.16	1.18		.10	.11	.06	.59	.11	
TiO_2	2.03	1.97	1.62		.82	1.10	2.52	.97	1.40	
P_2O_5	.31	.21	.15				.41	.25	.22	
MnO	.22	.26	.16		.08	.10	.25	.13	.23	.12
Cl	.01	.00	.01		.10	.03	nd	nd	.02	
F	.05	.07	.14				nd	nd	.18	
S		.09	.15		.03	.01	.13	nd	10	
Less 0 for F, S	.03	.06	12						12	
Total	99.91	100.03	99.82						99.98	

APPROXIMATE MODES

	SK2-7	SK4-6	FA-4	39-L	Am2*	Am3*	Am166	Am167
Hornblende	8.6	58.8	57.2		X	X	42.4	38.7
Clino- pyroxene	39.7	13.6	3.7		X		—	
Biotite		1.1	2.3			X	27.8	11.2
Andesine	22.8	25.3	18.2		X	X	28.6^{An40}	36.5
K-Feldspar	2.1		4.6		X			
Quartz	6.8	.6	6.2					6.8
Magnetite		.8	2.8				0.1	
Apatite		R						0.8
Sphene	4.1				X			0.7
Scapolite	12.3				X			4.7
Garnet	.2	.1						
Carbonate	1.6		2.0					

* See description of specimens.

DESCRIPTION OF SPECIMENS

SK2-7   Slightly calcareous amphibolite intermediate between diopsidic dolomite and Sk4-6 (in the adjoining column of this table). This amphibolite is formed by skarnlike replacement of a clean dolomite and is intimately intertongued with marble of the Grenville series, 5½ miles east of Gouverneur, Northwest Adirondack Mountains. Analyst J. J. Engel.

SK4-6   Amphibolite formed by interaction of fluids from granite upon clean dolomite of the "Grenville series," Gouverneur marble belt, 5½ miles east of Gouverneur, N.Y. Analyst J. J. Engel. This rock grades laterally into specimen SK2-7 which is interlayered with diopsidic, scapolitic marble.

FA-4   Feather amphibolite formed by skarnlike replacement of marble of the "Grenville series," 4½ miles south of Bancroft, Ontario, on Highway 62. Here as in many exposures in the area, all transitions exist between amphibolites and marble. Analyst J. J. Engel.

39-L   Amphibolitic layer in anorthosite. Garnet-bearing andesine (Ab^{67})-hornblende rock, 2½ miles south of Schroon River quarry on main highway to Schroon Lake,

Am2
Am3
Am166
Am167
AAm1.

Paradox quadrangle, Adirondacks, N.Y. Interpreted by R. Balk as a gabbro-amphibolite flow layer; by A. F. Buddington as a metamorphosed layer of "Grenville" origin. Analyst, R. B. Ellestad. From Buddington, 1939, Table 10.

Am2 Amphibolite derived from marble of the "Grenville series" and interlayered with marble at Maxwell's Crossing, Haliburton-Bancroft areas, Ontario. From Adams and Barlow, 1910, p. 105, who report the rock is composed of hornblende, pyroxene, plagioclase, scapolite, and an "untwinned feldspar," with accessory sphene.

Am3 Amphibolite "end stage" in amphibolization of marble, Haliburton-Bancroft areas, Ontario. From Adams and Barlow, 1910, pp. 104–105. This amphibolite occurs as an inclusion in granite, in the same series of exposures as Am2. Mineral constituents: hornblende, plagioclase, and biotite.

Am166 Metagabbro amphibolite, at contact of metagabbro with granite 2½ miles east of St. Regis falls, Nicolville Quadrangle, Adirondacks, N.Y. Data from Buddington, 1939, pp. 186–187, Table 44.

Am167 Metagabbro amphibolite cut by a network of granite pegmatite, Russell Quadrangle, Northwest Adirondacks. Analyst A. Willman. Data from Buddington, 1939, pp. 186–187, Table 44.

AAm1. Analyses of a composite sample composed of 23 specimens of amphibolite, layers intercalated in the major paragneiss of the "Grenville series," Gouverneur Quadrangle, Northwest Adirondacks. These amphibolites of unknown origin (see Engel and Engel, 1953b, pp. 1059, 1071–1072, 1083), are dominantly hornblende-andesine rocks with subsidiary biotite, quartz, K-feldspar, and clino-pyroxene. Accessories include serecite, chlorite sphene, zoisite, and magnetite. Analyst J. J. Engel.

anorthosites or syenites in other Precambrian terranes of North America? Comparison of the observations of five of the most experienced students of things Grenville, Balk (1931), Buddington (1939, 1948), Hewitt (1953), Osborne (1936), and Morley Wilson (1925), suggests there is no concensus regarding the properties, origin, or age relations of the igneous-looking suites in the Grenville subprovince.

GRENVILLE SUBPROVINCE AND GRENVILLE OROGENIC BELT

Presumably it has been necessary to define "Grenville series" before defining and delimiting the "Grenville subprovince" as the area which contains Grenville series? Hard on the heels of these terms, a new one, namely, "Grenville orogenic belt" has appeared. Does the "Grenville orogenic belt" refer to that segment of the Grenville series (?) in the Grenville subprovince(?) which has been deformed?

My questions are of long duration but triggered perhaps by the recent speculations of McLaughlin (1954), who has proposed an extension of the "Grenville province" and the "Grenville orogenic belt." It would seem reasonable to ask what exactly is being extended?

McLaughlin states (1954, p. 287): "the [Grenville] province is approximately bounded on the southeast by the St. Lawrence River, but with the Adirondack Mountains as a southeastern outlier. The northwestern border of the province lies about 250 miles northwest of the St. Lawrence. This boundary is known as the Grenville front."

In McLaughlin's paper the Grenville series is not defined or discussed. The "Grenville orogenic belt" is referred to very briefly. McLaughlin comments (1954, p. 287):

The extent of the Grenville orogenic belt is not yet fully established. J. Tuzo Wilson (1948) suggested that it is truncated at the northeast—by the Labrador orogenic belt, but later he seems to have regarded the relationship as an open question (1949). The linears shown on his earlier map may allow the interpretation of the two belts as contemporaneous.

Following this, McLaughlin notes (p. 287):

To the southwest, the Grenville belt passes out of sight under early Paleozoic sediments north of Lake Ontario, with no indication of diminished intensity of folding, if we may judge from the intricate structure in the Hastings district (Thompson, 1943). It could be expected to continue for some hundreds of miles. Directly along its strike lie three anticlinal structures in the Paleozoic rocks. In order, from northeast to southwest, these are (1) the Findlay arch, (2) the Cincinnati dome, and (3) the Nashville Dome. It is suggested that these broad, gentle upwarps are the present surface expression of the Grenville mountain range.

Finally, McLaughlin uses these domes as stepping stones into the Ozarks, which he suggests are still a further extension of the Grenville orogenic belt, now displaced northward by a great fault up the Mississippi embayment.

These are large steps. They are noted here chiefly because of the concepts they involve. One critical concept is that of the extent of the

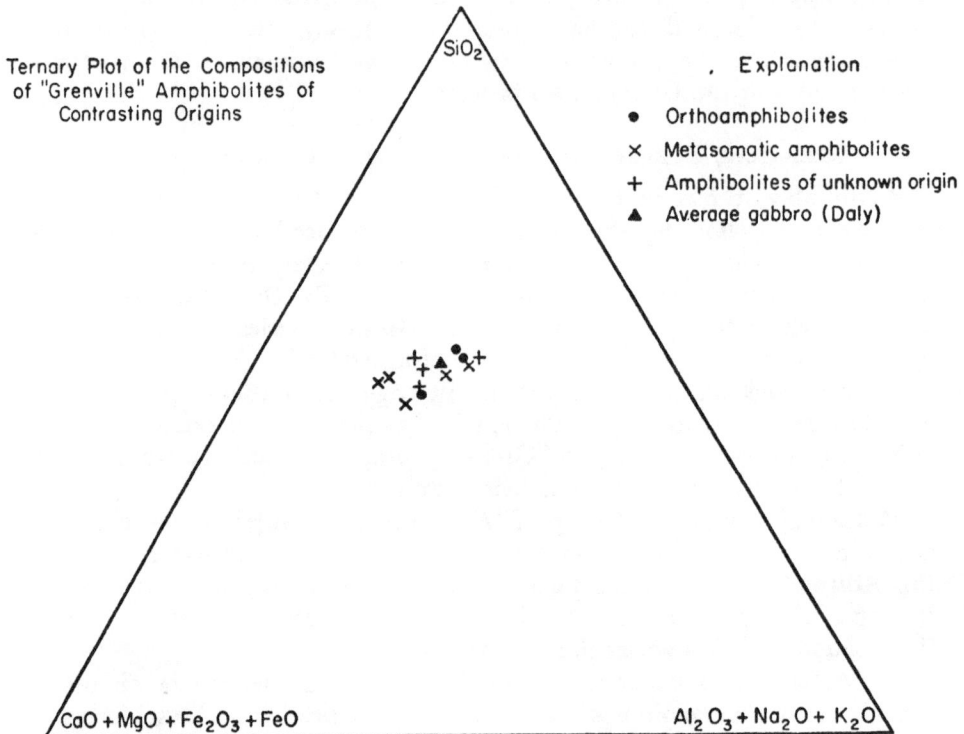

Ternary Plot of the Compositions of "Grenville" Amphibolites of Contrasting Origins

. Explanation

• Orthoamphibolites
× Metasomatic amphibolites
+ Amphibolites of unknown origin
▲ Average gabbro (Daly)

SiO₂

CaO + MgO + Fe₂O₃ + FeO

Al₂O₃ + Na₂O + K₂O

FIGURE 5. Ternary plot of the compositions of "Grenville" amphibolites of contrasting origins. The specimens represented include all of those whose analyses and modes are given in Table III. Three additional amphibolites are taken from Buddington (1939, specimens 55, 57 and 164-L, Table 44, p. 186).

Grenville subprovince. Another is that of a genetic association of Grenville subprovince and an *orogenic belt*. For it is neither a rock sequence nor a lithologic type that is correlated or extended some 1000 miles southwest into the Ozarks of Missouri, but certain geomorphic and structural(?) trends. Precambrian rocks are not exposed in either the Cincinnati or Nashville domes, and almost nothing is known of them from drilling. Some Precambrian is exposed in the Missouri Ozarks, however, and a great deal more of it is known from drilling and mine operations, especially along its southeastern flank. The rock types are in general Grenvillianous. Volcanics are especially abundant together with fairly massive granites and diabase dikes. Metasediments are very rare in the outcroppings, although some gneisses cut in drill holes appear to be of metasedimentary origin (J. S. Brown, personal communication). No marble or quartzites are known, nor are there definite ages established for the Ozark orogeny(ies).

McLaughlin had endorsed the suggestions of Adams and Barlow (1910, p. 16), and J. T. Wilson (1949; 1954, pp. 192–193) of a close genetic, temporal and spatial relationship between the Grenville sediments and environment of sedimentation, and the intrusive and orogenic cycle which follow. Is there really an invariant genetic relation between all Precambrian sedimentary sequences, associated igneous rocks, and tectonic axes as Wilson suggests? Pettijohn (1943, pp. 926, 960–962) concluded that the "Timiskaming" and "Keewatin" type association of greywacke, volcanics, pillow lavas, and lean siliceous iron formations, were the characteristic associations formed in the axial areas of evolving geosynclines. He reasoned that foreland areas of carbonate and quartzitic sedimentary rocks either never existed, or were destroyed by deep erosion.

In contrast, the characteristic lower Huronian (James, 1951) and Grenville carbonate-shale-quartzite sequences suggest to many a site at least intermediate between craton—foreland—(Kay, 1947) and geosynclinal axis (Pettijohn, 1943; Wilson, 1954), if not the craton itself (J. T. Wilson, 1954; see also the summary in Engel and Engel, p. 1037–1044, 1953a). The conception that the carbonate and well-sorted clastic segments of the Lower Huronian and Grenville rock types imply more profound residual weathering and perhaps recycling is a corollary of the ideas of relative stabilities of the respective sites of sedimentation. So, in a sense, are many workers' concepts of age relations between what has been referred to as (younger?) "Grenville type" and the (older?) "Timiskaming" and "Keewatin" types. Perhaps both the greywacke-volcanic and the limestone-quartzite-shale associations form at different times, in different sites in the axial areas of evolving geosynclines? But these are genetic concepts and we have not agreed whether or how they are to be written into our definitions.

Apropos the extension of the "Grenville orogenic belt," it is notable that in this symposium, Buddington has treated the New Jersey Precambrian as "Grenville." He is hardly alone in this assumption (see, for example, the summary in Engel and Engel, 1953a, pp. 1030–1031, 1043). The striking similarity of metasediments, igneous rocks, ore deposits, and deformative features in Vermont and New Jersey to these features in the Adirondacks and near Grenville Village have long been noted. Even the isotopic composition of ore lead from the New Jersey Precambrian approaches that at Tetrault, Quebec, much more than the ores of the Mississippi valley (Kulp, 1954; Nier, 1939). Hence an isotopic correlation of New Jersey and Quebec lead ore may be in order.[7]

It would seem that ground rules far more rigorous than McLaughlin has employed could apply in extending the Grenville subprovince to the Atlantic Coast. Certainly if the Ozarks are to be inducted into the Grenville subprovince, the Precambrian of New Jersey is a charter member. By recognizing New Jersey as an outlier of the Grenville subprovince, we obtain *Lebensraum* in which to build a really sizable Grenville subprovince to the southwest. Certainly almost any realistic definition of the "Grenville series" or "subprovince" that includes much of the southwestern one-third of the subprovince southeast of the Grenville front, must incorporate the Vermont and New Jersey outliers of Precambrian. And if this is done, the younger Appalachians, with their accordant tectonic axes, are neatly snuggled within the confines of the Grenville subprovince. But let us abandon this tantalizing expansion for the present.

Conclusion

Probably we all agree that attempts at lithologic correlations of rocks as injected and dynamothermally metamorphosed as those in the Grenville subprovince, are full of peril and possible misunderstandings. Leith cited many of the problems two decades ago (1934). As our errors of omission and commission have come home to roost and doubt has multiplied, we have blended the genetic with the descriptive, until seemingly, there are no recognized and functional standards of the term "Grenville series." Can we arrive at some working definitions that will enhance rather than inhibit progress and understanding? There are well-formulated and sound methods of separating descriptive and interpretative problems in sedimentary rocks (Schenck and Muller, 1941; Moore, 1947). Are these rules of correlation as practised by stratigraphers in unmetamorphosed fossiliferous sediments far too demanding and neither broad nor flexible enough to suit our needs? We are not always sure which rocks *are* of sedimentary origin as contrasted with those of igneous or hybrid origin. The various hypotheses of origin of the amphibolites, nepheline syenites, and the "granites" in the Grenville are painful reminders of this fact.

The metamorphic and igneous overprint not only obstructs our recognition of which rocks were sedimentary as contrasted to igneous or metasomatic types, but of course blurs the details that tell us specifically what kinds of sediments. Both amphibolites and the gneisses are examples

[7]The younger Ozark leads are highly Grenvillianous.

of this difficulty. Each is a major component of so-called Grenville series almost everywhere this term is applied. Yet the amphibolites could be (and in places seem to be) (1) thoroughly reconstituted tuffs, (2) argillaceous or ferriferrous carbonates, (3) basic volcanics, (4) diabase sills, as well as (5) metasomatic derivatives of relatively pure dolomites or limestones. The so-called "Grenville type" paragneisses may be in part tuffs, shales, argillaceous sandstones, arkoses, and greywackes. The correlations implied and the origin designated in any instance may be right or wrong, but in either event the implications in terms of environment of sedimentation and thus the crustal mechanics during sedimentation, are of enormous importance to our understanding of the evolution of North American continental and shield areas. This seems to be a critical aspect of our problem, as may be indicated by comparison of these two rock types, amphibolite and paragneiss, at the type locality with the same types in the northern and eastern parts of the province. The widespread references to the "type Grenville" as a sequence of limestone, quartzite, and normal shale is based upon the belief that the garnetiferous sillimanitic paragneiss is a "normal shale." Such a sequence if considered in the context of our contemporary ideas of sedimentation, as related to environment and tectonics, may be thought of as formed on a stable shelf or in an epicontinental sea. But suppose the paragneiss is in reality a metamorphosed tuff or subgreywacke, secondarily altered, perhaps, as some seem to have been, whereas the associated amphibolites are volcanics and mafic tuffs. This alternative association of limestones and quartzites with volcanics and tuff or greywackes could reflect a very different environment of sedimentation and related crustal behaviour.

Suggestions; Designations of Groups

Actually none of these problems preclude a distinction between descriptive and interpretative facets of the Grenville problem. Rodgers (1954, p. 655), in describing the methods of stratigraphers in sedimentary rocks, notes:

Stratigraphy has three tasks. First, it must subdivide and describe local stratigraphic sections. Second, it must attempt to show the time relations of these local sections to each other and to a standard geochronologic sequence understood by all stratigraphers; that is, it must *correlate* the local sections. Third and most important, it must *interpret* the local sections over a smaller or larger region, deducing the geologic history of the region as recorded by these sections.

The author would suggest that this is also the best way to attack the rocks often designated Grenville series, and indeed all Precambrian terranes.

The first step, the subdividing and describing of local sections, has been in progress for a century. Unfortunately, the second step also has been attempted, but upon little or no valid basis, or without careful recognition of the assumptions involved. Too often no clear distinction is made between the several objectives. For, as is widely recognized, the word "series" as well as "system" (larger) and "stage" and "zone" (smaller) are almost universally used for expressing time-stratigraphic units (Rodgers, 1954). Use of the term "Grenville series" away from the type locality is therefore quite misleading, impractical, and unrealistic.

The commonly accepted unit names employed for the subdivision and description of local sections—whose time and stratigraphic relations are unclear—are group, formation, member, and so on. These so-called rock-stratigraphic or lithogenetic units, especially the unit *group*, seem well adapted to the immediate tasks that confront us.

For example, group names—having no implication of geologic age or stratigraphic affinities—might be given to the better known, distinctive, metasedimentary sections heretofore called "Grenville series" in various parts of the "Grenville subprovince." For the present the number of groups might be restricted to those better known sections which seem to represent contrasting lithologic types. Seemingly transitional or intermediate *lithologic types* could be described as such, without any implication of geologic time or stratigraphic correlation until the interrelations of the groups are clarified by detailed mapping and age studies. Interpretations of the origin and evolution of each rock type, and of the history of the region as inferred from study of the various sections, could form a complementary but not confusing aspect of the work.

Perhaps a clearer understanding of the above is possible by use of several specific examples. Although these examples are based upon a review of all the published map sheets and reports of areas in the "Grenville sub-province," they are merely to be regarded as highly tentative suggestions. Any formal designations of stratigraphic or lithologic types or groups should come from those national, provincial, and state groups especially concerned with geologic mapping in the region.

The available data on the metasedimentary-volcanic(?) sections in the Grenville subprovince suggest the relative abundance and distributions of rock types shown in Table I. These admittedly crude designations of the averages and range in abundances of the common lithologic types are obtained from published maps and reports. In each map area the amount of rock designated or implied as metasedimentary together with amphibolite and volcanics was calculated. Wherever possible the breakdown into the amounts of marble, quartzite, paragneiss, and so on, were made. These breakdowns were weighted according to size of the map area and averaged with data from nearby map areas. The results undoubtedly can be refined and improved by many geologists, especially the staffs of the Quebec and Ontario Department of Mines and the Geological Survey of Canada. But the tabulated figures will serve as a first approximation and an example.

As indicated in Figures 1 and 2, and earlier (Engel and Engel, 1953a), three distinctive sections appear to exist: (1) at and near the type locality, (2) in the Hastings region (Bancroft-Kingston region?), and (3) at most mapped areas in the northeast one-half to one-third of the subprovince (Parent-Moisie River region?).

The type locality (Grenville series per se) seems to be dominated by marble and potassic paragneiss together with quartzite and amphibolite. The amphibolites are largely or wholly of unknown origin. No definitely identifiable volcanics or conglomerates are reported.

In contrast to the type locality the sections in the Hastings region contain conglomerate, some sodic (as well as potassic?) paragneisses, identifiable volcanics with abundant marble. Quartzite and amphibolite also are common, but the distinctive elements there are conglomerate, some sodic gneiss, and identifiable volcanics in a marble-rich sequence. It seems possible that the percentage of marble in the section exceeds that at the type locality, but this is not clearly demonstrated.

Most of the sections in the northeast one-half to one-third of the province appear to be dominated by a sodic paragneiss and amphibolite of obscure origin. Quartzites seem to be somewhat impure, marbles and conglomerates are rare or absent. Volcanics may have been abundant but rarely are identifiable. The characteristic features of these sections are therefore (1) abundance of sodic gneisses and amphibolite, (2) scarcity of both carbonate sediments and presumably conglomerate.

Possibly at least two or many groups are needed in this vast region, one say on the lower Saguenay, another northwest of St. Félicien, and so on.

Compared to these sections in Canada, the Precambrian sections in intervening areas and in the Adirondacks, Vermont, and New Jersey, appear to be transitional in lithologic type. Thus in the Adirondacks, marble and a sodic paragneiss are the major metasedimentary types, with amphibolite, quartzite, and subordinate units of potassic paragneiss. This "section" seems transitional in its lithologic features between those features in the Hastings area and at the type locality. So indeed do the Vermont and New Jersey sections.

The group sections probably should be named to facilitate description and discussion. The names might designate areas where the section is well exposed and mapped. Thus, the sections in the Hastings area might be designated the Madoc group or Kaladar group, etc.; sections in the northeast one-third or one-half of the region, the Saguenay, Moisie, St. Félicien group, etc.; whereas the rocks at the type locality could be referred to as the Grenville or Chatcham group. The author wishes to emphasize that these are being used as examples and not as the newly proposed group names.

The fact that one or more of the groups may include unconformities, rocks of quite different age or origin, does not lessen the value of the group names. Stratigraphic and structural reconstructions, correlations and interpretations of the origin of the component rocks must be encouraged and accelerated, but genetic concepts should be distinguished carefully from the designation and descriptions of groups. Rocks in intermediate areas might be referred to as like one or another group or as exhibiting composite features.

Actually each suggested group appears to involve the oldest metasedimentary rocks in the respective areas.[8] These metasediments with or without associated volcanics are intruded by one or more igneous types such as the well-known anorthosites, gabbros, syenites, and granites. Consequently, the various igneous or igneous-looking complexes form a sort of datum or minimal age designation for the groups. The many dates in absolute years placed upon pegmatites, granites, and pitchblende-bearing veins (Collins et al., 1954, Tilton, et al., 1954; Hurley, 1951, Tables 2 and 3; Ellsworth, 1932, pp. 102–106; Ellsworth and Osborne, 1934; Nier, 1939, p. 159; Kulp, 1954, p. 1275) in the region indicate an age of 800 million to 1.2 billion years for the most recent metamorphism and igneous intrusion. Conceivably dates may be affixed to earlier intrusives or metamorphic periods at various places, and in this way the minimum age of the youngest associated metasediments may be established. Actual dating or bracketing of the periods of sedimentation or of specific units within the groups seems far away (Engel, 1955).

If the happy day comes when the actual time of deposition of the Precambrian sedimentary rocks can be calculated, the term "Grenville series" may become applicable to rocks away from the type locality. Until then, there is much fun to be had in describing the rocks and in unravelling their possible origins and interrelations as these bear upon the history of the region.

REFERENCES

ADAMS, F. D. (1896). Report on the geology of a portion of the Laurentian area lying to the north of the Island of Montreal (Quebec); Geol. Surv., Canada, Ann. Rept. 1895, 184 pp.

—————— (1909). On the origin of the amphibolites of the Laurentian area of Canada; Jour. Geol., vol. 17, pp. 1–18.

ADAMS, F. D. and BARLOW, A. E. (1910). Geology of the Haliburton and Bancroft areas, Province of Ontario: Geol. Surv., Canada, Memoir 6, 419 pp.

BALK, ROBERT. (1931). Structural geology of the Adirondack anorthosite: A structural study of the problem of magmatic differentiation; Min. petrog. Mitt., Neue Folge, Band 41, Heft 3–6, pp. 308–434.

BUDDINGTON, A. F. (1939). Adirondack igneous rocks and their metamorphism; Geol. Soc. America, Memoir 7, 354 pp.

—————— (1948). Origin of granitic rocks of the northwest Adirondacks; Geol. Soc. Amer., Mem. 28, pp. 21–43.

CLAVEAU, JACQUES (1950). North shore of the St. Lawrence from Aguanish to Washicoutai Bay, Saguenay County; Quebec Bur. Mines, Geol. Rept. 43, 40 pp.

COLLINS, C. B., FARQUHAR, R. M. and RUSSELL, R. D. (1954). Isotopic constitution of radiogenic leads and the measurement of geological time: Bull. Geol. Soc. Amer., vol. 65, pp. 1–22.

DELAND, A. N. (1953). Surprise Lake Area; Quebec Dept. Mines, Prelim. Rept., 11 pp.

DE LA RUE, E. A. (1948). Nominingue and Sicotte map areas, Labelle and Gatineau counties; Quebec Dept. Mines, Geol. Rept. 23, pp. 1–59.

ELLSWORTH, H. V. (1932). Rare element minerals of Canada; Geol. Surv., Canada, Econ. Geology ser. 11, pp. 102–106.

ELLSWORTH, H. V. and OSBORNE, F. F. (1934). Uranite from Lac Pied des Monts, Saguenay district, Quebec; Am. Mineral., vol. 19, pp. 421–429.

ENGEL, A. E. J. and ENGEL, CELESTE G. (1953a). Grenville series in the northwest

[8]A possible exception is the region of Quebec northwest of St. Félicien.

Adirondacks Mountains; Part I, General features of the Grenville series; Bull. Geol. Soc. Amer., vol. 64, pp. 1013–1047.

——— (1953b). Grenville series in the Northwest Adirondack Mountains; Part II, Origin and metamorphism of the major paragneiss; Bull. Geol. Soc. Amer., vol. 64, pp. 1049–1097.

FAESSLER, C. (1936). Suzor-Letondal map area, parts of the counties of Laviolette, Saint-Maurice and Abitibi; Quebec Bur. Mines, Ann. Rept., pt. 13, pp. 25–36.

——— (1948). Simon Lake area, Papineau County; Quebec Bur. Mines, Geol. Surveys Br., Geol. Rept. 33, pp. 1–29.

GILBERT, J. E. (1952). Rohault Area, Abitibi-East and Roberval counties; Quebec Dept. Mines, Prelim. Rept., 10 pp.

GILLIES, N. B. (1952). Canimiti River Area, Pontiac County, Quebec; Quebec Dept. Mines, Geol. Rept. 52, 46 pp.

GREIG, E. W. (1945). Matamec Lake map area, Saguenay County; Quebec Dept. Mines, Geol. Survey, Geol. Rept. 22, p. 1–28.

GRENIER, PAUL E. (1953). Gamache area, Abitibi-East County; Quebec Dept. Mines, Prelim. Rept., 10 pp.

HARDING, W. D. (1944). Geology of Kaladar and Kennebec townships; Ont. Dept. Mines, 51st Ann. Rept., pt. 4, 1952.

HEWITT, D. F. (1953). Geology of the Brudenell-Raglan area; Ont. Dept. Mines, 62nd Ann. Rept., vol. 62, pt. 5, 123 pp.

HURLEY, PATRICK M. (1951). Alpha ionization damage as a cause of low helium ratios; U.S. Office Naval Research Tech. Rept., 10 pp.

JOHNSTON, W. G. Q. (1954). Geology of the Timiskaming-Grenville contact southeast of Lake Temagami, Northern Ontario, Canada: Bull. Geol. Soc. Amer., vol. 65, pp. 1047–1074.

KAY, MARSHALL (1947). Geosynclinal nomenclature and the craton; Bull. Amer. Assoc. Pet. Geol., vol. 31, pp. 1289–1293.

KULP, J. LAURENCE and BATE, GEORGE L. (1954). Variation in the isotopic composition of common lead; Bull. Geol. Soc. Amer., vol. 65, pp. 1275.

LEITH, C. K. (1934). The Precambrian: Proc. Geol. Soc. Amer., 1933, pp. 151–180.

LOGAN, W. E. (1863). Geological Survey Canada. Report of progress from its commencement to 1863, 938 pages.

MATTHEW, G. F. (1909). As cited in Precambrian Geology of North America; U.S. Geol. Survey Bull. 360.

MCLAUGHLIN, DEAN B. (1954). Suggested extension of the Grenville Orogenic Belt and the Grenville Front; Science, vol. 120, no. 3112, pp. 287–289.

MEEN, V. B. (1944). Geology of the Grimsthorpe-Barrie area; Ont. Dept. Mines, 51st Ann. Rept., pt. 4, 1942, pp. 1–50.

MILLER, W. G. and KNIGHT, C. W. (1914). The precambrian geology of southeastern Ontario: Ont. Bur. Mines, vol. 22, pt. 2, 151 pp.

MOORE, R. C. (1947) Note 2—Nature and classes of stratigraphic units; Bull. Amer. Assoc. Pet. Geol., vol. 31, pp. 519–528.

NIER, A. E. (1939). The isotopic constitution of radiogenic leads and the measurements of geological time, II; Phys. Review, vol. 55, ser. 2, pp. 153–163.

OSANN, A. (1902). Notes on certain Archean rocks of the Ottawa valley; Geol. Surv., Canada, Rept. Prog. 1876–77, pp. 1–84–0.

OSBORNE, F. F. (1936). Intrusives of part of the Laurentian complex in Quebec; Am. Jour. Sci., 5th ser., vol. 32, pp. 407–434.

PETTIJOHN, F. J. (1943). Archean sedimentation; Bull. Geol. Soc. Amer., vol. 54, pp. 925–972.

——— (1949). Sedimentary rocks. New York, Harper, 514 pp.

RETTY, J. A. (1944). Lower Romaine River area, Saguenay County; Quebec Bur. Mines, Div. Geol. Surveys, Geol. Rept. 19, pp. 1–31.

RODGERS, JOHN (1954). Nature, usage, and nomenclature of stratigraphic units; Bull. Amer. Assoc. Pet. Geol., vol. 38, pp. 655–659.

ROSS, S. H. (1949). Geological reconnaissance of Peribonca River; Quebec Dept. Mines, Geol. Surveys Br., Geol. Rept. 39, 20 pages.

SATTERLY, J. (1942). Mineral occurrences in Parry Sound District; Ont. Dept. Mines, 41st Ann. Rept., pt. 2, pp. 1–80.

———— (1943). Mineral occurrences in the Haliburton area; Ont. Dept. Mines, 52nd Ann. Rept., pt. 2, pp. 1–96.

SCHENCK, H. G. and MULLER, S. W. (1941). Stratigraphic terminology; Bull. Geol. Soc. Amer., vol. 52, pp. 1419–1426.

THOMSON, J. E. (1943). Mineral occurrences in the north Hastings area; Ont. Dept. Mines, 52nd Ann. Rept., pt. 3, pp. 1–80.

TILTON, G. R. *et al.* (1954). The isotopic composition and distribution of lead, uranium, and thorium in a precambrian granite; U.S. Atomic Energy Commission (AECU-2840), 27 pp.

TODD, E. W. (1928). Kirkland Lake gold area: A detailed study of the central ore zone and vicinity; Ont. Dept. Mines, Ann. Rept., vol. 37, pt. 2. p. 20.

VAN HISE, CHARLES RICHARD, and LEITH, CHARLES KENNETH (1909). Pre-Cambrian geology of North America; U.S. Geol. Survey Bull. 360, chapt. 5, pp. 448–483.

VENNOR, H. G. (1878). Explorations in Renfrew, Pontiac, and Ottawa counties, with notes on apatite, plumbago, and iron ores of Ottawa County; Geol. Surv., Canada, Rept. Prog. 1876–77, pp. 244–301.

WADDINGTON, G. W. (1950). Limestone deposits of the Mingan Islands area; Quebec Bur. Mines, Div. Geol. Surveys, Geol. Rept. 42, pt. 2, pp. 1–48.

WILSON, J. T. (1949). Some major structures of the Canadian shield; Can. Min. Metal. Bull., vol. 42, no. 451, pp. 547–554.

———— (1954). The development and structure of the crust; in The Earth as a Planet, ed. G. P. Kuiper, University of Chicago Press, pp. 138–214.

WILSON, M. E. (1924). Arnprior-Quyon and Maniwaki areas, Ontario and Quebec; Geol. Surv., Canada, Mem. 136, pp. 1–152.

———— (1925). The Grenville Precambrian subprovince; Jour. Geol., vol. 36, pp. 389–407.

———— (1933). The Claire River syncline; Trans. Royal Soc. Can., Series III, vol. 27, Sec. 4, pp. 7–11.

———— (1939). The Canadian shield; in Geology of North America, vol. 1 of Geologie der Erde, ed. Erich Krenkel, pp. 232–311. Gebrüder Bornträger.

DISCUSSION

J. E. THOMSON

I am delighted with Engel's talk and I am in complete agreement with him. I believe that we must start from scratch and build up new descriptions of the entire Precambrian succession. I should like to add that I am also in entire agreement with what Dr. J. E. Gill said yesterday about Precambrian nomenclature (Trans. Roy. Soc. Can., Ser. III, vol. XLIX, 1955, Section IV, pp. 25–29).

J. E. GILL

I agree with Thomson and I can see that there is a new light dawning in Precambrian nomenclature. While these revisions are underway I should like to see the name Keweenawan thrown out as well except for use in the type area because there are diabases of several different ages. I believe that I have been accused of introducing new names for provinces unnecessarily. The first comprehensive divisions of the Shield boundaries were placed at areas of ignorance, due to surface cover or lack of mapping. Following the publication of the new geological map of Canada in 1948, I attempted a revision, based on structural trends, and two new names were introduced to avoid confusion. I used the name "Grenville province" instead of "Grenville subprovince" because the area is not a subdivision of any larger province from a structural viewpoint. Furthermore it represents a large part of the Shield.

D. F. HEWITT

I must admit I like the term "subprovince" though it is often hard to make generalizations.

D. R. DERRY

I like the terms "Grenville front, Grenville area, and Grenville sediments." We must have some terms to use in discussion and everyone understands the meaning of these. Let us make the broad distinctions first and then proceed to work out the details.

M. WALTON

I agree with Derry and also to some extent with Engel but there is a danger of throwing out the baby with the bath. I like the term "Grenville subprovince" which can be defined as an area in which there are pegmatites about 1000 million years old. There is a tendency to confuse this with the stratigraphic meaning of the word Grenville. It seems to me to be all right to call a major tectonic unit "Grenville." There are similar tectonic units and names in use elsewhere. It is not helpful to think of the Grenville area as one of Grenville sequence sediments. I like the terms "front" and "subprovince."

A. E. ENGEL

I agree that the terms "front" and "subprovince" are useful but let us define them precisely. If anyone says there is no confusion let him read all the Grenville literature and he will find that the confusion in the use of terms is enough to drive him nearly crazy. If anyone wishes to continue an old name, would he please define how he means it to be used in clear, basic English. The recent paper in *Science* by Dean McLaughlin is an example of how confusion arises, because he has used the Nashville and Cincinnati domes as stepping stones to carry the Grenville province all the way to the Ozarks. I am glad that geology is getting so much help from other scientists. This it certainly needs. I should like to say that I am grateful to the Toronto group for their age determinations.

M. E. WILSON

The history of the name "subprovince" is this: Van Hise and Leith in Monograph 52 of the United States Geological Survey divided the Lake Superior Region into north and south subprovinces. Age determinations by geophysical methods are inexact because they determine the age of intrusive rocks that may be hundreds of millions of years younger than the sediments they intrude. For this reason, continuity of outcrop is still by far the best way of correlating Precambrian rocks. However, because of the barriers of Hudson Bay, the Hudson Bay Lowlands, Lake Superior and areas of granitic rocks, correlation in the Canadian Shield by means of continuous outcrop is impossible. I, therefore, in 1928 when writing the chapter on the Canadian Shield for the Bornträger *Geologie der Erde*, in the volume on the Geology of North America (not published until 1939) divided the Shield into provinces and subprovinces. For the approximate Precambrian part of the St. Lawrence drainage basin, the name "St. Lawrence province" was proposed. Within the St. Lawrence province, two additional subprovinces, Timiskaming and Grenville, were added to those of Van Hise and Leith. Whatever the merit of Dr. Gill's suggestion that the St. Lawrence province be divided into Superior and Grenville provinces, the division would introduce so much confusion in names, that the change can scarcely be warranted.

CORRELATION OF RIGID UNITS, TYPES OF FOLDS, AND LINEATION IN A GRENVILLE BELT

A. F. Buddington

REGIONAL AND DETAILED STUDIES of the geology of the Precambrian rocks of the Adirondacks, New York, and of the New Jersey Highlands during the last fifteen years by members of the United States Geological Survey have afforded data which permit some attempts at new large-scale structural synthesis and certain reinterpretations of detailed structure. The rocks involved are similar in character to those of the Grenville series in Canada and have commonly been thought to belong with them.

There are large gaps in the geologic mapping and in essential data so that any discussion of regional structure can only be of a preliminary nature.

The author is indebted to B. F. Leonard who co-operated in geologic mapping of the St. Lawrence County magnetite district and is co-author of a report on that district (in preparation) from which Figures 1 and 2 are taken. The writer is also indebted to D. R. Baker who co-operated in geologic mapping of the Beaver Lake anticline in New Jersey (Fig. 6).

Three different series of major igneous intrusives are found in the metasedimentary series in the Adirondacks. They are represented successively by series of rocks related to gabbroic anorthosite and anorthosite, pyroxene syenite and pyroxene quartz syenite, and hornblende microperthite granite. There are at least five periods of intrusion of gabbroic or diabasic rocks as dikes or sheets but these are not pertinent to the present discussion of structure. The first two major series are inferred to have been emplaced during periods of non-orogenic activity and the third during a period of orogeny. All three series represent abnormal igneous types compared with the usual major post-Precambrian intrusives; yet they are the major igneous rocks in a great belt several hundred miles wide that extends from Labrador at least to Virginia. The nature of the deformation of the whole complex during the last epoch of Precambrian orogeny has been perhaps somewhat unique, because of the presence of such great relatively rigid units as the anorthosite bodies; thick sheets of orthogneiss, generally quartz syenitic in composition, also acted as relatively rigid buttresses when folded into anticlinal units.

Salic magmas appeared during both the second and the third major periods of intrusion. The earlier salic magmas were of quartz syenitic composition. Where they are in thick sheets, the rocks are differentiated, probably in place, into pyroxene syenitic facies below and pyroxene quartz syenitic or hornblende granitic facies above, with transitional facies between (Buddington, 1948, pp. 24–30). The younger salic magmas were of granitic composition and are represented by hornblende microperthite granite and biotite alaskite.

The anorthositic series of rocks are inferred to have been derived from magma. The recent data of Yoder (1955, pp. 106–107) on the system $H_2O\text{-}CaO.Al_2O_3.2SiO_2\text{-}CaO.MgO.2SiO_2$ are revolutionary in showing that the eutectic ratio for $CaO.Al_2O_3.2SiO_2\text{-}CaO.MgO.2SiO_2$ is shifted from 42 per cent anorthite 58 per cent diopside to 73 per cent anorthite 27 per cent diopside in a liquid with 9 per cent H_2O at 5000 atmospheres pressure and 1095°C. The cotectic ratio of labradorite to mafics in the presence of a high percentage of H_2O may be expected to be considerably higher than for anorthite and hence comparable to the composition of the gabbroic anorthosite magma ($\pm$ 85 per cent feldspar) inferred (Buddington, 1939, pp. 234–237) to have been the primary magma source for the rocks of the anorthositic series in the Adirondacks. Very large masses of gabbroic magma are known to have been emplaced ,as great lopoliths (Duluth, Bushveld, etc.) in relatively flat lying rocks during Precambrian time. Gabbroic anorthositic magma can be thought of as the water-rich equivalent of these emplaced under similar structural conditions but yielding a raised instead of a depressed roof consistent with a much lighter density than normal gabbroic magma. Such a feldspathic-rich magma would yield a small amount of late stage normal gabbroic magma through escape of H_2O without much temperature change. The plagioclase of gabbro and anorthosite would thus be of similar composition, as is the case. The relatively high alumina content of the clinopyroxenes of anorthosite is consistent with the hypothesis of crystallization from an anorthositic magma. The extensive local metasomatic replacements at the borders of the anorthosite bodies and the occurrence of hematite on a substantial scale as exsolved intergrowths in accessory ilmenite in the anorthosite in the place of the less oxidized compound magnetite are all consistent with the postulate of a water-rich anorthositic magma.*

All the major rocks of the Adirondacks, with the exception of part of the younger hornblende granites, the core of the anorthosite masses, and sporadic local lenses, have been intensely deformed and recrystallized or reconstituted to gneisses. It is our fundamental assumption, however, that for the most part the secondary planar structure of the orthogneisses is

*A mechanism for producing an H_2O-rich magma from normal basaltic magma may be based on the principle recently elaborated by Kennedy (Crust of the Earth, Special Paper no. 62, Geol. Soc. Am., pp. 489–503, 1955).

conformable with a primary flow structure or with gravity differentiated stratiform layers and may be used to infer gross structural forms, and that arrangements of the major lithologic units of the metasediments may also permit us reasonable inferences as to the nature of their fold structures.

Except to a subordinate extent there is apparent conformity of rock units that is thought to be the result of primary conformity. Pseudoconformity of tectonic origin is present, but a consistent picture is obtained if a primary conformity is assumed to be predominant.

The metasedimentary rocks have undergone intensive deformation in the solid state. Engel and Engel (1953, p. 1029) have studied this in detail and concluded: "In spite of the tendency for many thin zones and some major units to flow or disperse, the bulk of the major rock motions and thus most dominant planes of secondary shear appear to have followed the sedimentary fabric. . . . Because sedimentary bedding remained the dominant guide to the shearing and penetrative movements in the Grenville series, the most prominent foliation and lithologic layering induced by metamorphism is, in most areas, a quasi-bedding foliation."

Other basic assumptions are that the anorthosites are products of differentiation of a volatile-rich magma of gabbroic anorthosite composition which was emplaced as very large bodies with a multidomical surface or in small masses as sheets and lenses essentially conformable with sediments having an initial gentle structure; that the masses of quartz syenitic magma were emplaced as sheets more or less conformable with the bedding of only gently folded or flat lying sediments and that the younger granite magmas were intruded contemporaneous with a period of orogeny as essentially conformable sheets and phacoliths. The emplacement of the granite magmas in large part as phacoliths in the "Highlands" belt of New Jersey and New York has also been shown by Lowe (1950), Hotz (1952), and Sims (1954). Cross-cutting relationships, however, were an accompaniment of the emplacement of the salic magmas.

WEDGE STRUCTURE, RIGID UNITS, AND OVERTURNED ISOCLINAL FOLDS

A wedge structure in the northwest Adirondacks has been previously described (Buddington, 1939, p. 241) but more data are now available.

The relationships between bodies of anorthosite and of early orthogneiss, which are inferred to have acted as relatively rigid units, to belts of overturned isoclinal folds, and zones of open to tight vertical folds in part of the Adirondacks are shown in Figure 1.

Inspection of the map shows a broad wedge structure between the granite orthogneiss of the Alexandria mass on the northwest and the Lowville dome and the anticline of the Stark complex on the southeast. The wedge structure is asymmetrical with a narrow belt ($2\frac{1}{4}$–4 miles wide) of isoclinal folds overturned towards the Alexandria mass in the northwest, a very broad belt (15–20 miles wide) in the southeast with isoclinal folds over-

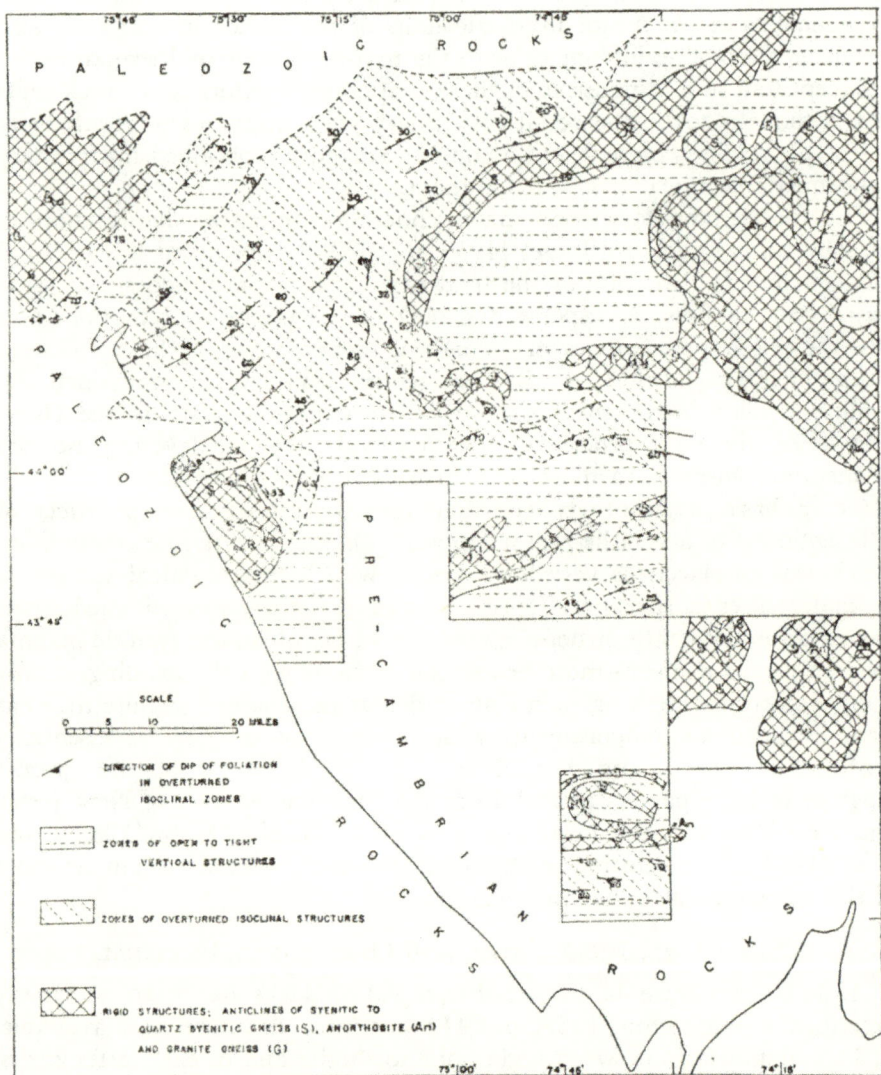

FIGURE 1. Relations between zones of overturned isoclinal folds and more rigid elements in mapped areas of the western Adirondacks. *Aa*—anticline of alaskite gneiss, Alexandria Bay area; *Ab*—anticline of pyroxene syenite gneiss, Arab Mtn. area, Tupper Lake quadrangle; *It*—pyroxene syenite gneiss of Inlet area. Cranberry Lake quadrangle: *Le*—anticlinorium of pyroxene syenite gneiss and hornblende quartz syenite gneiss of Lowville dome; *Po*—pyroxene quartz syenite gneiss and granite gneiss of Piseco Lake dome; *St*—anticline of hornblende granite gneiss of Stark complex; *Tl*—anticline of syenite gneiss, Twitchell Lake area, Big Moose quadrangle.

turned towards the southeast, and a narrow belt (6 miles wide) with folds having more or less vertical axial planes between the border zones.

It will also be noted that a belt of isoclinal folds overturned towards the units of early syenitic and granitic gneiss forms a complete arc around the Stark complex, Inlet mass, and Arab Mtn. anticline. Again, in the southern part of the map, there is a belt of isoclinal folds overturned towards the Piseco dome of quartz syenite gneiss and granite gneiss. At the north end of the Lowville dome the border facies of the rocks on opposite sides of the dome itself are isoclinally overturned, each of them towards the core of the dome.

Another feature is the roughly circular area within the framework of the orthogneiss of the Stark, Inlet, and Arab Mtn. units and the anorthosite with its overlying fringe of syenite orthogneiss. Within this area there are many folds, but overturned isoclinal structures are rare. It is as though the rocks of this area had been somewhat protected from the most intense deformation by the relatively rigid rim of early orthogneiss and anorthosite. There is the additional possibility that the area is underlain at depth by a mass of anorthosite which has served as a rigid plate.

Another wedge structure is exemplified by the relationships between the anticlinorial orthogneisses of the Twitchell Lake area (Big Moose quadrangle) on the northwest and the composite mass of anorthosite and syenite-quartz syenite gneiss on the southeast. The width of the wedge between the two units is a little over 15 miles. Folds are isoclinally overturned towards the southeast on the southeast and towards the northwest on the northwest with a central zone of moderately open to tight folds with limbs having opposed dips.

Balk (1932, p. 44) shows a general cross-section across the Newcomb quadrangle (NW corner is 74°15′ Long., 44°00′ Lat.) which also represents a wedge-shaped structure. His data on foliation show steep isoclinal folds slightly overturned towards the north on the north border and a uniform northerly dip on the south border with vertical foliation between.

In general, then, we may say that all the belts of isoclinal folds are arranged in such a fashion that they are peripheral to units of anorthosite and early orthogneiss, or of orthogneiss alone, and are overturned towards them. This leads to the inference that the anorthosite and orthogneiss units have acted as relatively rigid elements during the deformation in which the overturning of the folds was effected. All the relatively rigid units of early syenite, quartz syenite, and granite orthogneiss are anticlinal or anticlinorial structures formed by the tight folding of thick stratiform sheets of igneous rock with concomitant crushing and variable degree of recrystallization in the process of solid flow. The development of these orthogneiss units preceded the period of later emplacement of the syntectonic younger granites which to a variable degree were themselves subjected to deformation in the solid state.

LARGE-SCALE RELATIONSHIPS OF LINEATION

The term lineation as used here refers almost wholly to the dimensional elongation of individual minerals and of mineral aggregates. To a minor extent, the orientation of crumples in paragneisses has been used in areas where these are thought to be parallel to mineral elongation.

The term "subparallel" will be used for lineation whose angle of rake in the plane of foliation is within 30° of the strike of the foliation, and the term "subperpendicular" will be applied to lineation whose angle of rake in the plane of foliation is within 30° of the dip of the foliation. The subparallel lineation corresponds to the "a" lineation and the subperpendicular lineation to the "b" lineation of some authors (Cloos, 1946). However, owing to the different definitions for the terms "a" and "b" lineation in the literature, they will not be used here.

All the masses of orthogneiss of the relatively rigid units shown in Figure 1 have, in so far as is known, a lineation subparallel to the major axes of folds, with rare local exceptions. By contrast the lineation is subperpendicular to the major fold axes in zones of overturned isoclinal folds adjacent to the Stark, Inlet, Arab Mtn., Lowville, and Alexandria units and subparallel to minor folds. Such zones of subperpendicular lineation may be up to twelve miles wide, as they are northwest of the overturned Stark complex (Fig. 2), or coincident with the whole belt of isoclinal structures, as they are along the southeast border of the Alexandria mass (Fig. 1).

The rocks in the central part of the wedge (Fig. 1) between the Alexandria mass and the Stark and Lowville units have a lineation subparallel to the axes of major folds. This central part of subparallel lineation includes both the belt of moderate to tight folds and a wide belt (up to ten miles) to the southeast in which the folds are isoclinally overturned to the southeast. Lineation subperpendicular to the major fold axes is thus confined almost wholly to belts of strongly overturned isoclinal folds, but lineation subparallel to major fold axes may also occur in large belts of overturned isoclinal folds.

The lineation of the rocks inside the arc formed by the Stark, Inlet, and Arab Mtn. units is in general parallel to this arc and subparallel to the fold axes which also are conformable with the arc. The area adjacent to the main anorthosite massif, however, has not been studied in detail.

Cleaves Rogers (personal communication) states that the lineation in gneisses is subperpendicular to the major fold axes in the southeastern part of the southeastern belt of overturned isoclinal folds southeast of the Twitchell Lake anticlinorial unit but is subparallel to the fold axes in the central part of the wedge structure to the northwest.

Figure 2 shows several characteristic structural relationships. On this map the term "mixed gneiss" includes a great variety of gneisses formed from metasedimentary rocks by reconstitution, migmatization, or modification by partial metasomatism, together with thin intrusive sheets and lenses

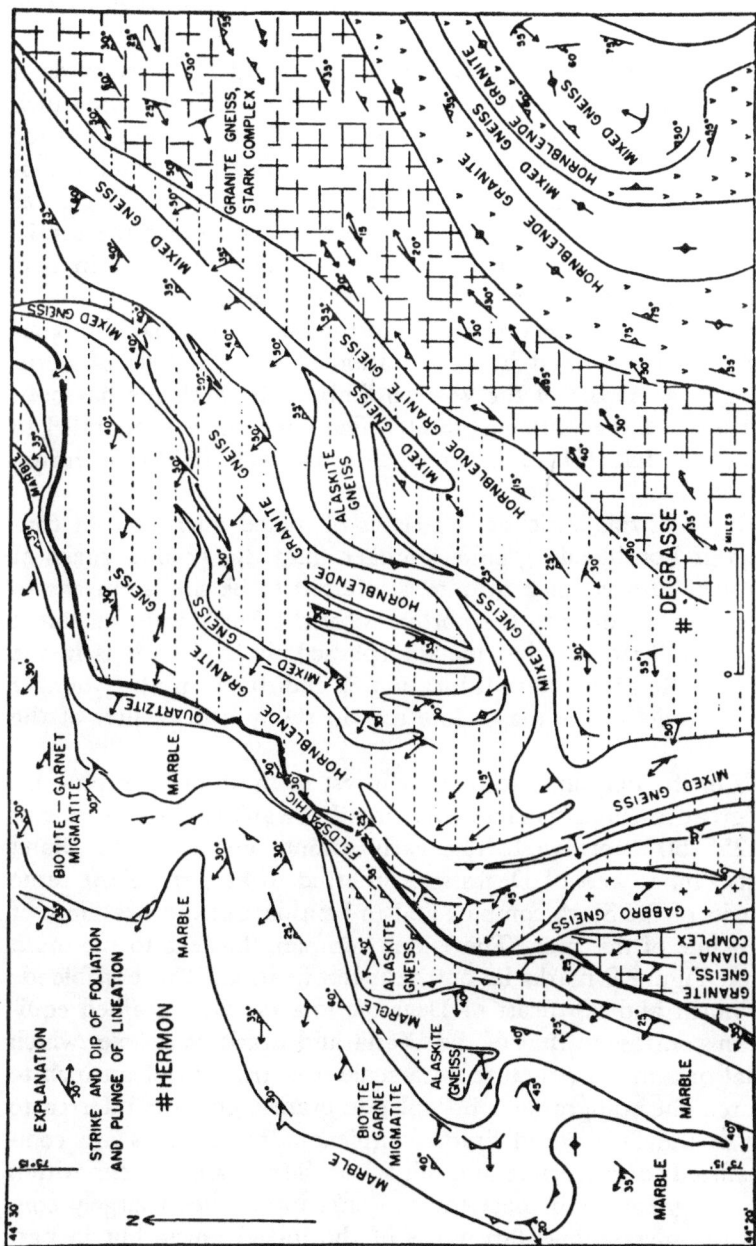

FIGURE 2. Details of overturned isoclinal syncline on northwest flank of relatively rigid unit of Stark complex. Lineation is subperpendicular to trend of major fold axes, Northwest Adirondacks. Heavy line marks boundary between belt on northwest in which rocks of the Grenville series predominate and belt on southeast in which igneous rocks and ortho-gneisses are predominant.

of granitic material. Part of the gneiss is a sillimanitic microcline-rich quartz-feldspar rock formed by replacement and modification of a biotite quartz plagioclase gneiss. A little marble or its skarn equivalent is also present.

The heavy line about half-way between Degrasse and Hermon marks a boundary between an area 24–30 miles wide on the northwest, in which rocks of the Grenville series form about three-quarters of the total, and the main area of igneous or metaigneous rocks to the southeast, in which orthogneisses and igneous rocks form about 85 per cent of the total. There has been strong differential movement along this boundary. Along the southeast side, ultramylonite is common in the gneisses. The rocks are in part intensely crackled, and locally there are chloritic slickensides.

A major structural unit is the orthogneiss of the Stark complex. This consists of a tight anticline formed by deformation of a thick sheet of hornblende granite gneiss. A part of the western limb of the anticline has been cut out by intrusion of the younger granite. The lineation is subparallel to the major axis of the fold. This mass acted as a relatively rigid unit during later deformation.

To the southeast of this anticline is a series of mixed gneisses with conformable sheets of hornblende granite younger than the granite gneiss of the Stark complex. This younger granite is inferred to be similar in age to the hornblende granite gneiss to the northwest of the Stark complex but it is much less recrystallized and less metamorphosed situated as it is in the lee of the early orthogneiss unit. Lineation is indistinct in the younger granite and subparallel to the major fold axis in the mixed gneisses of the syncline.

All the rocks of the map area to the northwest of the Stark complex are involved in a great isoclinal syncline overturned toward the southeast and with dips of 25°–50° northwest. The early granite gneiss of the Diana complex about 5 miles west of Degrasse is inferred to be part of the same sheet as the gneiss of the Stark complex and to be a part of the west limb of an isoclinal syncline of this rock. The Diana complex thickens to the south and the part on Figure 2 marks its northern termination. The hornblende granite gneiss north and northeast of Degrasse is a medium grained equigranular rock in contrast to that of the Diana and Stark complexes which has a phacoidal or almond structure on a scale of a fraction of an inch to an inch or more. The equigranular hornblende granite gneiss is inferred to be younger than the Diana and Stark complexes. Its feldspars are completely recrystallized to microcline and oligoclase. The granite gneiss within the main area of igneous and metaigneous rocks forms sheets largely conformable with the metasedimentary rocks of the mixed gneiss but in part cross-cutting them.

The metasedimentary rocks northwest of the main area of igneous and metaigneous rocks show extreme thickening and thinning in consequence of plastic flow in the solid state. The biotite-garnet migmatite gneiss may be terminated at the south end, perhaps by thinning in consequence of being pulled apart from its disrupted extension to the southwest.

The marble is from a mile in width to zero. There has been a little thickening of the feldspathic quartzite on the crest of the minor fold about 6.5 miles east of Hermon. The variation in thickness of the feldspathic quartzite layer is in part a consequence of the variable extent to which the granite gneiss underlying it contains included layers. Two small phacoliths of alaskite have been emplaced within the minor folds several miles south-southeast of Hermon.

The lineation of all the rocks northwest of the Stark complex is sub-perpendicular to the major fold axes and subparallel to the minor folds.

Subparallel and Subperpendicular Lineation in Diana and Loon Lake Complexes

The change from lineation subparallel to the major fold axis to lineation subperpendicular is well illustrated in the Diana and Loon Lake ortho-gneiss complexes. In each area the change coincides with the place where the fold becomes intensely overturned.

Diana Complex

Figure 3 shows the structure of part of the Diana complex, which is a complexly folded differentiated sheet of average quartz syenitic composition. The sheet formed from a quartz syenitic magma that differentiated into a pyroxene syenite with local shonkinite lenses rich in ilmenite and magnetite in the lower part, a hornblende granite in the upper part, and transitional hornblende syenite and quartz syenite between. The sheet has a maximum thickness of about 3.5 miles southeast of Harrisville and thins to the north-east and southwest. Subsequent to emplacement the sheet was intensely folded and the rock itself strongly deformed and in large part recrystallized. The northwest border facies in the southwest portion of the map area (Fig. 3, south of Natural Bridge) has undergone severe cataclastic crushing and in considerable part has been changed to a mylonite. Farther northeast this border facies is very largely a mortar gneiss with a cataclastic mortar. The southeastern portion of the mass through Naumburg, Aldrich School, and Kalurah was deformed at a higher temperature than the northwest part (Buddington, 1952) so that primary microperthite feldspars have been differentiated into recrystallized independent grains of potash feldspar and plagioclase accompanied by some porphyroblastic development of horn-blende. The texture is granoblastic with variable amounts of relic porphyro-clasts of feldspar. A younger granite mass has cut into the complex along the southeast border.

The general trend of the major folds is northeast. The areas of pyroxene syenite gneiss may be considered as the lowest stratigraphic horizon of the sheet and the hornblende syenite, quartz syenite, and granite gneisses as the upper stratigraphic portion. The rocks of the southwestern part shown

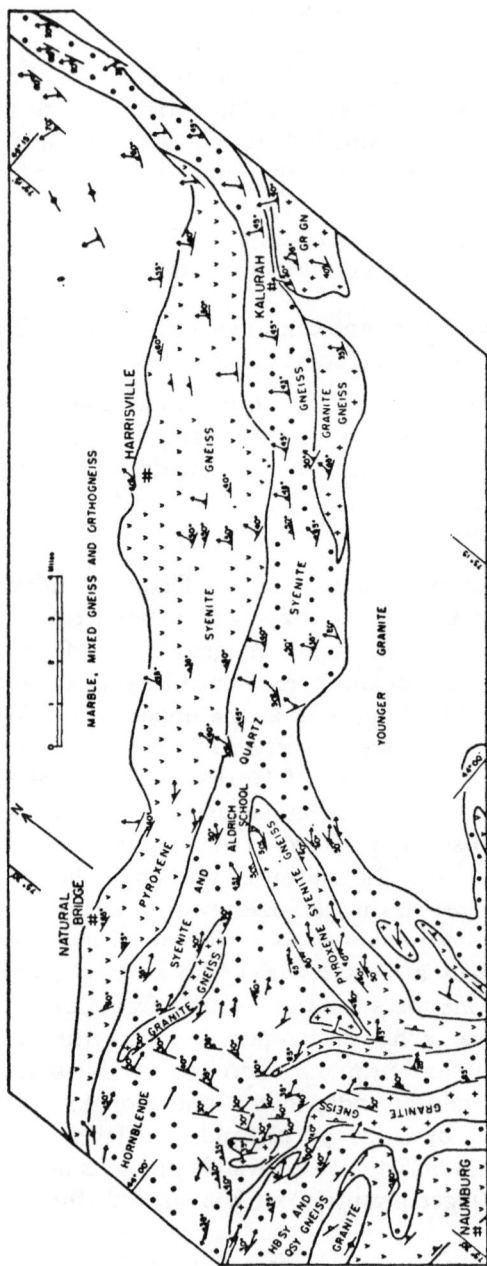

FIGURE 3. Part of Lowville dome and limb of isoclinal syncline formed from differentiated stratiform sheet of the Diana complex. Lineation is subparallel to major fold axes on dome and is subperpendicular to major fold axis on overturned limb. Northwest Adirondacks. Symbol for strike and dip of foliation and plunge of lineation same as in Figure 2.

on the map between Naumburg, Aldrich School, and Natural Bridge are
on the northeast-plunging nose of a great dome (Lowville dome, Fig. 1)
of complexly folded gneiss. Two north-northeast striking anticlinal crumples
on the nose of the great dome are shown south-southwest of Aldrich School
and a few miles south-southwest of Natural Bridge. The Lowville dome as
a whole, as shown to the southwest beyond the area of the map, is at least
15 miles across. South of Natural Bridge the rocks are isoclinally overturned
to the south whereas southwest of Aldrich School the rocks are isoclinally
overturned towards the northwest.

To the northeast of the Lowville dome the sheet has a monoclinal dip of
about 50° to the northwest and the succession of the rocks shows that it is
overturned to the southeast, and is the overturned northwest limb of an
isoclinal syncline. The northern termination of the Diana complex is shown
in the western part, southern border of Figure 2. The western limb of the
anticline of the granite gneiss, shown as part of the Stark complex in the
eastern part of Figure 2, is inferred to be the eastern limb of an isoclinal
syncline of which the Diana complex is the western limb.

The lineation shows two highly contrasted relationships to the trend of
the major fold structures. In the northeastern part of the map (Fig. 3)
from Aldrich School to beyond Kalurah the lineation has a strike of north-
northwest to northwest, or about subperpendicular to the trend of the rocks,
which is about N 50° E from Natural Bridge to Kalurah. In the area between Naumburg, Aldrich School, and Natural Bridge the lineation strikes
about east-northeast, and is thus subparallel to the inferred long axis of the
Lowville dome which has about a N 50° E strike. The lineation is definitely
of post-consolidation origin for it also affects hypersthene diabase dikes with
chilled borders which transect the structure of the Diana complex. The
east-northeast lineation is on the northeast-plunging nose of the Lowville
dome and is thus subparallel to trend of the major fold. The change to a
subperpendicular type occurs where the broad Lowville dome of complex
folds passes into a strongly overturned simple monoclinal limb of an isoclinal
syncline. A reasonable hypothesis would be that the lineation in the early
stages of deformation was developed subparallel to the trend of the major
fold throughout, but that in the later stages of deformation the broad Lowville dome acted as a relatively rigid unit permitting accentuation of the
original linear structure whereas the simple limb was strongly overturned
with intense forward motion to the southeast and the development of a new
lineation in conformity with the direction of principal movement and subperpendicular to the major fold axis.

Loon Lake Complex

The Loon Lake quartz syenite complex is shown in Figure 4. This is a
modified map taken from a report on the Saranac Lake quadrangle (Buddington, 1953). The southwestern part of the map shows a broad local
depression with a synclinorium of younger gneisses in a part of the northern

border portion of the main Adirondack anorthosite mass. A great sheet of orthogneiss overlies the anorthosite. It consists of rock with an over-all composition of quartz syenite and a medium-grained equigranular texture. The sheet is differentiated and metamorphosed so that the upper part is a pyroxene-hornblende quartz syenite gneiss and the lower part is a garnetiferous pyroxene syenite gneiss. North of Saranac Lake there is a syncline of metasediments with a sheet of granite gneiss. The metagabbro mixed with granite gneiss may be in a synclinal structure or may be a lenticular mass. The granite gneisses are younger than the syenite-quartz syenite gneiss. The extreme southeast corner of the map shows a small portion of another syncline with a sheet of differentiated quartz syenitic material overlying the

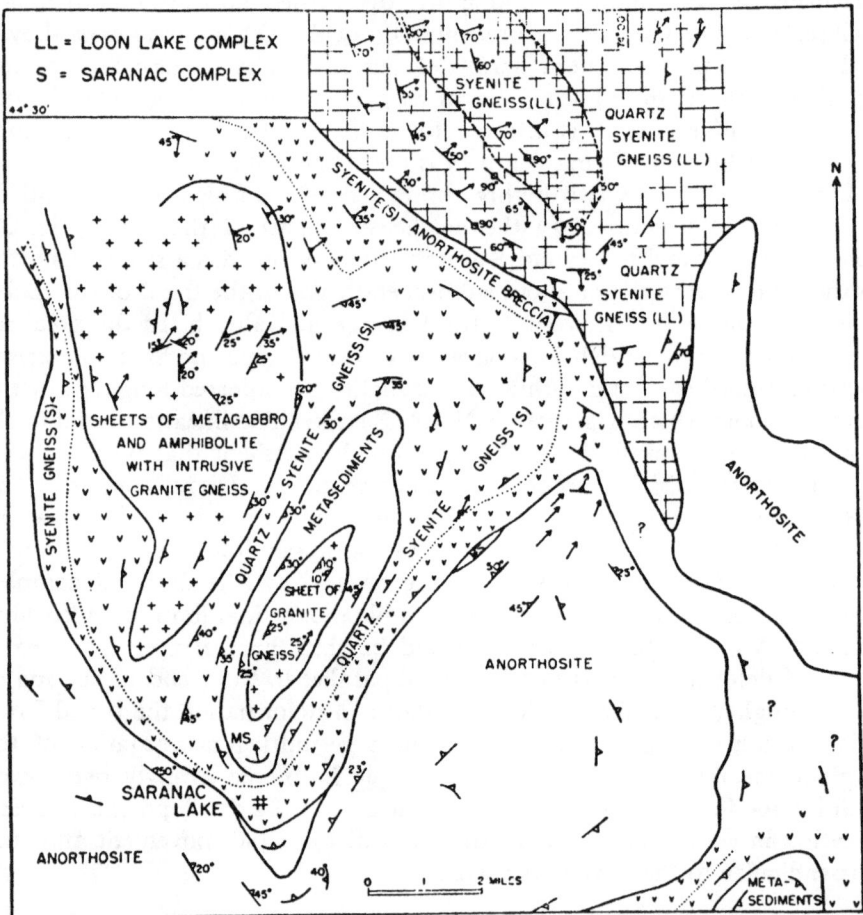

FIGURE 4. Area in central Adirondacks. Lineation is subparallel to major fold axes (of rocks) in synclinorium within part of the relatively rigid anorthosite massif; lineation is subperpendicular to major fold axes (northern part of map) in strongly overturned isoclines. Planar foliation indistinct or absent where lineation alone is shown without foliation symbol. Symbol for strike and dip of foliation and plunge of lineation same as in Figure 2.

anorthosite and with metasedimentary rocks in the trough. The limbs of all the folds mentioned have moderate dips and the linear structure is subparallel to the major fold axes.

The northeastern part of the map shows an anticline of porphyroclastic quartz syenite gneiss plunging to the south, with an intrusion breccia of syenite into anorthosite on the southwest flank and anorthosite on the east side of the nose. The linear structure in this anticline is subparallel to the strike of the fold on its nose where the limbs have opposed dips. To the northwest, however, the fold passes into an isocline overturned to the southwest and in this part of the fold the lineation is subperpendicular to the trend of the major fold. The rocks are intensely deformed where the overturning has taken place. The change in strike of the lineation from subparallel to subperpendicular coincides with the change from a tight asymmetrical fold with opposed dips on the limbs to a strongly overturned isoclinal type.

The sheet of porphyroclastic or phacoidal quartz-syenite gneiss is separate from the differentiated even-grained quartz-syenite gneiss to the southwest but the relative age relations are unknown.

Plunging Noses with Subparallel Lineation Only

Wherever the lineation is subparallel to the fold axes in folded sheets or phacoliths of orthogneiss, it is characteristic for planar foliation to be indistinct or indeterminate and linear structure only to be present in the crestal zone of plunging noses of anticlines or in the plunging keel of tight synclines. This relationship is shown in Figure 4 in the plunging nose of anorthosite gneiss at the north end of the local anorthosite dome east of Saranac Lake. It is also exemplified in the Piseco dome and Beaver Lake anticline shown in Figures 5 and 6.

Piseco Dome

The rocks and structure of the Piseco dome (Fig. 1) have been described in excellent detail by Cannon (1937, pp. 60–63 and maps 1–3). A map modified from his results is given in Figure 5, and the following summary is based on his descriptions. The rocks in the anticline are, from oldest to youngest, metasedimentary rocks with amphibolite, hornblende-pyroxene quartz-syenite gneiss, and granite gneiss. The quartz-syenite gneiss has a granoblastic structure, with locally a mosaic so fine-grained as to approach mylonite. The feldspar is in part relic microperthite. There is local sporadic development of porphyroblastic hornblende and sparse garnet. The rock of the granite gneiss also has a granoblastic texture with leaf quartz. The size of the grains increases from south to north. The quartz-syenite gneiss does not thicken on the noses of the fold as does the granite gneiss. It is therefore inferred that the quartz-syenite gneiss represents deformed sills whereas the granite gneiss is the product of initial emplacement as phacoliths and later deformation. The secondary foliation is assumed to be superimposed con-

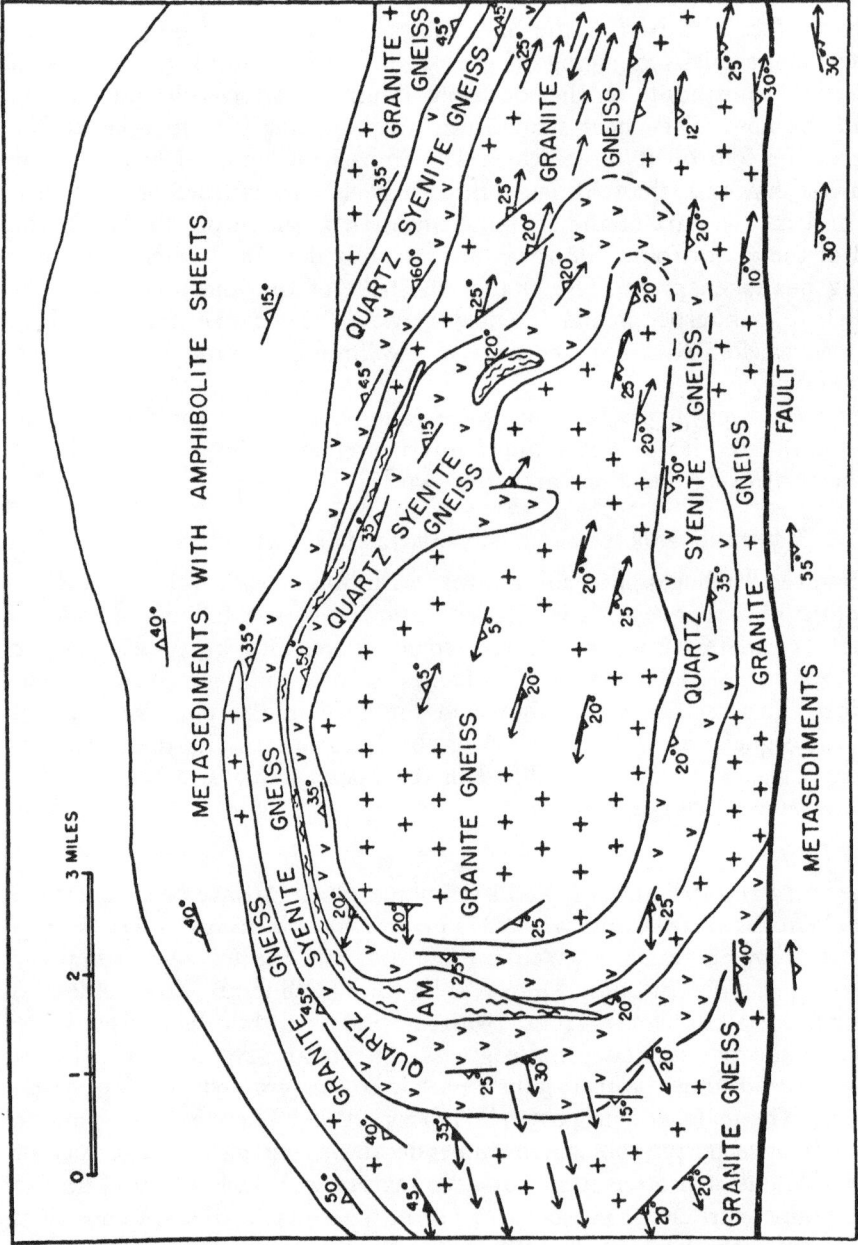

FIGURE 5. Map of Piseco dome. Lineation is parallel to major axis of folded quartz syenite gneiss sheet and of granite phacolith. Planar foliation indistinct or absent where lineation alone is shown without foliation symbol. Southwestern Adirondacks. (After R. S. Cannon.)

formably, except at the pinched ends of the folds, on a primary foliation which was in turn conformable with the borders against the country rock. Linear structures are essentially parallel to the axis of the dome, both in strike and in plunge. They are especially well developed on the plunging noses of the fold where a planar foliation is weak. Where the planar foliation is discernible in the granite gneiss at the ends of the dome, it is commonly irregular and folded in minor crenulations parallel to the major fold. The fold has only moderate dips on the limbs, and the north flank has a somewhat steeper dip than the south. The axial plane is therefore assumed to dip south, which is in accord with the greater intensity of granulation on the south.

Beaver Lake Anticline, New Jersey

A map of the Beaver Lake anticline, which is exposed in the Franklin Furnace and Hamburg 07′ 30″ quadrangles in New Jersey, is shown in Figure 6. The belt termed paragneiss includes metasedimentary rocks and their migmatitic and modified facies. The migmatitic facies are in considerable part sillimanitic. The granitized metasediments are mostly fine grained quartz-microcline gneisses, in part sillimanitic, inferred to be largely granitized biotite quartz plagioclase gneiss. The metasedimentary rocks and amphibolite (AM) are the oldest gneisses.

The core of the anticline is a quartz-oligoclase gneiss of uncertain origin. This gneiss extends for many miles to the southwest so that the portion shown on the map, except for the extreme southwest part, is the nose of a northeast-plunging anticline. The syenite gneiss has been emplaced as a syntectonic phacolith with the usual great thickening on the nose of the fold. The granite on the northeast nose of the fold is a younger phacolith. The syenitic rock has been deformed in the solid state and partly recrystallized but the granite is a microperthite variety which only locally shows crushing and recrystallization. As shown by the map, planar foliation is well developed only on the flanks of the fold except at the southwest and where it extends throughout the fold. The zone at the southwest border of the map where planar foliation occurs throughout the rock also characterizes the anticline for many miles to the southwest. To the northeast, however, in the crestal zone of the nose of the plunging anticline only lineation can be determined in the field. This applies not only to the gneisses but also to the granite, where the linear structure is fundamentally a product of fluid flow; orientation of inequidimensional grains during flow acted as a control for subsequent magmatic crystallization and growth. A planar foliation does occur in the base of the syenite gneiss on the nose of the anticline where it has been controlled by drag against the underlying amphibolite. This planar foliation in the syenite gneiss is inferred to represent a secondary foliation of solid flow superimposed on a similar gneissoid structure of fluid flow. The plunge of minor fold axes on the southeast flank of the anticline averages 45°–55° NE whereas the plunge of the major

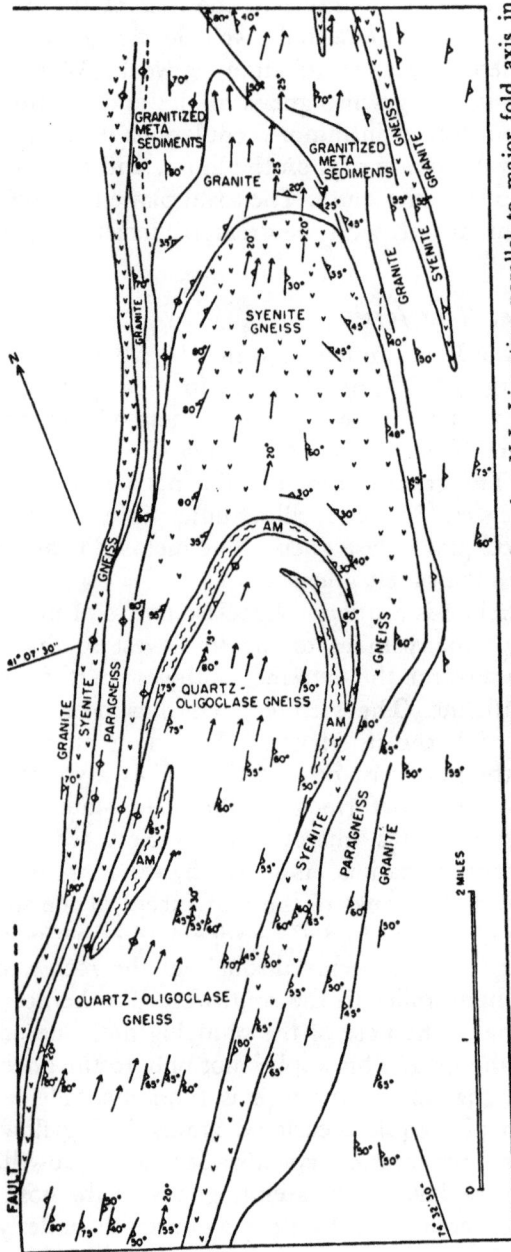

FIGURE 6. Beaver Lake anticline, Franklin and Hamburg quadrangles, N.J. Lineation parallel to major fold axis in gneiss and in granite phacolith and developed to exclusion of planar foliation on crest of plunging nose of anticline. Planar foliation indistinct or absent where lineation alone is shown without foliation symbol. Symbol for strike and dip of foliation and plunge of lineation same as in Figure 2.

anticlinal axis averages about 25° NNE. The anticline is asymmetrical with a steep slightly overturned limb on the northwest and a moderately steep dip on the southeast.

SOME FABRICS OF PHACOLITHS IN THE NORTHWEST GRENVILLE WEDGE

Many granite bodies occur in the wedge of the Grenville series southeast of the Alexandria mass (Fig. 1). Buddington (1929) infers that in large part they have been emplaced as syntectonic phacoliths; individual bodies have been mapped by Martin (1916) and Dietrich (1954).

The structure and intensity of deformation of the phacoliths is consistent with the nature of the deformation of the belt in which they occur.

The Fish Creek phacolith lies in the northwestern border portion of the wedge of the Grenville series where the rocks are in isoclinal folds overturned towards the northwest. It has been studied by the writer and described in detail by Dietrich (1954, pp. 513–531). The rock of the phacolith is in general completely recrystallized but has local undeformed facies or marked cataclastic facies in unsystematic spatial arrangement. The phacolith, according to Dietrich, occupies the trough of a syncline that plunges steeply northeast; the axial plane dips steeply (70°–80°) southeast. There are numerous amphibolite slabs included in the gneiss. In part these are oriented athwart the prevailing gneissic structure and the foliation is parallel to the longer dimensions of the amphibolite plates. Locally the amphibolite is crumpled as a result of shear folding and the foliation is across the layers, parallel to the foliation of the enclosing gneiss. In part there is local deviation of the gneissic structure of the granite to parallel the foliation of the amphibolite layers immediately adjacent to them. The foliation of the gneiss is in an arc conformable with its border at the southwest end.

Within the central zone of tight vertical to moderately open folds the granite of the phacoliths is in substantial part microperthite and quartz, uncrushed and unrecrystallized, with a gneissoid structure arising initially from fluid flow. The gneissoid structure is conformable with the wall rocks throughout some of the phacoliths (cf. Gouverneur phacolith, Buddington, 1929, p. 55). In the others, the gneissoid structure conforms in general with the wall rock, but throughout the core is of a planar type oriented subparallel to the axial plane of the fold of which it is a part. Included thin layers of amphibolite on the plunging noses of the phacoliths commonly have a crumpled structure (cf. Hyde School phacolith, Buddington 1934, p. 241, fig. 44) conformable with the gneissoid structure of the enclosing granite.

The Canton and Pyrites phacoliths are representative of those found in the southeast border belt of the wedge characterized by isoclinal folds strongly overturned to the southeast. They have been described by Martin (1916) and Buddington (1929).

The Canton compound phacolith occurs in an isoclinal fold strongly overturned to the east-southeast. The granite as a whole is intensely deformed, crushed, and gneissic. There is a strong lineation striking west to west-northwest about subperpendicular to the trend of the major fold. At the broad southwest end of the fold where it rakes gently south-southwest the foliation swings around in an arc generally conformable with the environing country rock and with thin included amphibolite layers. Locally, however, the foliation of amphibolite layers may strike west-northwest, whereas the foliation of the enclosing granite strikes north-northeast.

At the northeast end of the mass north of Pyrites there are chevron shear folds in the granite gneiss, clearly shown by thin amphibolite layers (Martin, 1916, pp. 98–99 and Pl. 19) whose axial planes are parallel to the general schistosity.

It may also be noted that the two alaskite phacoliths a few miles south-southeast of Hermon (Fig. 2) are emplaced concordantly in the crest of a minor fold and have a lineation subperpendicular to their trend.

There is a coarse porphyroclastic granite gneiss which occurs for the most part associated with biotite-quartz-plagioclase gneiss and probably in large part as a product of its granitization. This gneiss has a phacolithic development in the sense that it is of syntectonic emplacement and more or less conformable with the fold structures which may be either anticlines or synclines. The phacoliths emplaced as magma, however, are alaskitic and occur almost exclusively on anticlinal structures, within a thick marble formation or along thin marble layers within paragneiss. The phacoliths have varied shapes. A skeleton type of phacolith may consist of two lenses on opposite limbs of an anticlinal nose. Another type may form a crescent on one nose of an anticline with an extension along one limb or symmetrically with extensions along both limbs. In other types the phacolith may come into the crest of the fold or on both noses of doubly plunging anticlines.

Most of the phacoliths have been subjected to intense deformation and some flow in the solid state. We must therefore consider the problem whether part or all of the thickening of the gneiss of the phacoliths on the noses of the folds is a consequence of plastic flow. None of our data seem to be consistent with such an hypothesis and some are definitely inconsistent. The quartz-syenite-gneiss sheets of the Piseco dome (Fig. 5) are in a belt of intense deformation, yet show no significant local thickening or thinning in consequence of solid flow. The feldspathic quartzite along the border of the main igneous complex shows some thickening on minor folds but not of an order of magnitude which would be essential to explain the usual amount of thickening on the noses of phacoliths. Finally, the granite of gneissoid phacoliths carries a microperthite feldspar and a magnetite with ±6 per cent exsolved ilmenite whereas the gneissic granite carries microcline and plagioclase as separate grains and an ilmenite-poor magnetite. These phenomena are consistent with the first type being a product of magmatic crystallization and the second of intense deformation in the solid state, yet both have similar gross forms.

LINEATION OF GNEISSES OF NEW JERSEY HIGHLANDS

It has long been recognized that the lineation in the gneisses of the New Jersey Highlands and in the extension of these gneisses to the northeast in New York and to the southwest in Pennsylvania is almost exclusively to the northeast subparallel to the trend and plunge of the fold axes and to the trend of the rocks. The plunge is usually relatively low (10°–30°) though plunges of 35°–60° are present in local areas. The gneisses generally strike northeast and dip moderately to steeply southeast. Many folds have been recognized. Most are asymmetrical, being slightly overturned to the northwest or isoclinally overturned towards the northwest, and plunge to the northeast. The Beaver Lake anticline (Fig. 6) is representative. Only locally has the lineation in the gneisses of the Highlands been found to be subparallel to the dip. In such cases the lineation is often related to strongly overturned isoclinal folds. The geology of several areas within the New Jersey, New York, and Pennsylvanian Highlands has been studied in detail, yet to the writer's knowledge no great fold has yet been found with a plunge to the southwest. Since the total length of the belt of gneisses is about 150 miles this presents a problem of major magnitude. No one has yet proposed an hypothesis, based on substantiating data, to explain so great a uniformity in direction of plunge of folds and lineation, *if* this actually is the situation in the Highlands. Our data are too inadequate, however, to be sure that this really is the case.

The lineation has been demonstrated to have served as a controlling element for several successive events at one locality, the Hibernia mine in the Dover district, N.J. Study of several long diamond drill holes a maximum of 2700 feet apart leads to the conclusion that paramphibolite layers, sheets of quartz oligoclase granite, albite granite pegmatite lenses. and long lath-shaped ore bodies of replacement magnetite ore all have an elongation which parallels the lineation.

CONCLUSIONS

The purpose of this paper has been primarily to present certain field-scale structure fabric patterns and certain relationships between them which characterize the Precambrian rocks of the Adirondacks and of the New Jersey Highlands.

Two great wedge structures have been described for the Adirondacks, and these lead to the inference that certain anticlinal blocks of anorthosite or of early orthogneiss have acted during deformation as relatively rigid blocks compressing the rocks between them and resulting in isoclinal overturning towards them in the border zones.

The data show that lineation subperpendicular to major fold axes in the gneisses is in effect exclusively correlated with strongly overturned isoclinal folds especially in zones adjacent to the relatively rigid border units. This relationship is inferred to be a consequence of relatively strong "forward motion" as has been emphasized in general by Cloos (1946, p. 30). There are, however, some phenomena related to the minor folds in the Adirondacks which could be interpreted in terms of a northeast movement of rocks on the northwest relative to those on the southeast (Engel, in preparation). Furthermore, lineation subparallel to major fold axes is also common in overturned isoclinal folds especially towards the central zones of the wedges. Thus although it cannot be conclusively affirmed that lineation subperpendicular to major fold axes in the Adirondack gneisses is genetically related to relatively strong forward motion in the principal direction of movement the field data are consistent with and on the whole favour such an interpretation.

REFERENCES

BALK, ROBERT (1932). Geology of the Newcomb quadrangle; N.Y. State Mus. Bull. 290.

BUDDINGTON. A. F. (1929). Granite phacoliths and their contact zones in the northwest Adirondacks; N.Y. State Mus. Bull. 281, pp. 51–107.

———— (1934). Geology and mineral resources of the Hammond, Antwerp and Lowville quadrangles; N.Y. State Mus. Bull. 296.

———— (1939). Adirondack igneous rocks and their metamorphism; Geol. Soc. Amer., Mem. 7.

———— (1948). Origin of granitic rocks of the northwest Adirondacks; Geol. Soc. Amer., Mem. 28, pp. 21–43.

———— (1952). Chemical petrology of some metamorphosed gabbroic, syenitic and quartz syenitic rocks; Am. Jour. Sci., Bowen volume, pp. 37–84.

———— (1953). Geology of the Saranac quadrangle; N.Y. State Mus. Bull. 346.

CANNON, R. S. (1937). Geology of the Piseco Lake quadrangle; N.Y. State Mus. Bull. 312.

CLOOS, E. (1946). Lineation; Geol. Soc. Amer., Mem. 18.

DIETRICH, R. V. (1954). Fish Creek phacolith, northwestern New York; Am. Jour. Sci., vol. 252, p. 513–531.

ENGEL, A. E. J. (1949). Studies of cleavage in the metasedimentary rocks of the northwest Adirondack mountains, N.Y.; Trans. Am. Geophys. Union, vol. 30, pp. 767–784.

———— Geology and talc deposits of the Balmat-Edwards District, N.Y.; U.S. Geol. Survey, report in preparation.

ENGEL, A. E. J. and CELESTE G. (1953). Grenville series in the northwest Adirondack Mountains, New York; Pt. I, General features of the Grenville Series, Pt. II, Origin and metamorphism of the major paragneiss; Bull. Geol. Soc. Am., vol. 64, pp. 1013–1098.

HOTZ, P. E. (1953). Magnetite deposits of the Sterling Lake, N.Y.–Ringwood, N.J. area; U.S. Geol. Survey Bull., 982–F.

LOWE, K. E. (1950). Storm King granite at Bear Mountain, New York; Bull. Geol. Soc. Am., vol. 61, pp. 137–190.

MARTIN, J. C. (1916). The Precambrian rocks of the Canton quadrangle; N.Y. State Mus. Bull. 185.

POSTEL, A. W. (1952). Geology of Clinton County magnetite district, N.Y.; U.S. Geol. Survey, Prof. Paper 237.

SIMS. P. K. (1954). Geology of the Dover magnetite district, Morris County, N.J.; U.S. Geol. Survey Bull. 982–G.

YODER, H. S. (1955). The system diopside-anorthite-water; Ann. Rept. of Director of the Geophysical Laboratory, Carnegie Institution, for the year 1953–1954.

DISCUSSION

A. E. ENGEL

It seems to me that Buddington's work in the Adirondacks is one of the outstanding examples in the world of effort by a single individual but I think that two kinds of lineation and folding must be distinguished. In the first kind competent units retain their shape and plastic units flow like diapirs giving rise to rotation lineation. The other kind is cross-folding. G. S. Brown and I have worked over some areas in the Adirondacks in which both occur and where there is cross-folding a beta lineation has been superimposed on an earlier alpha lineation.

A. F. BUDDINGTON

I agree 100 per cent with Engel on the different possibilities for formation of lineation. The reason why I cannot agree on the development of lineation sub-perpendicular to the major fold axes by differential movement perpendicular to the lineation, is that in the cases cited this would have to occur primarily only where the folds were strongly overturned.

A. E. ENGEL

I believe that that is a relic structure.

F. F. OSBORNE

Is the lineation the $\beta \perp \beta'$ of Sander?

A. F. BUDDINGTON

I have avoided using a and β because I prefer the terms sub-parallel and sub-perpendicular in view of confused usage.

F. F. OSBORNE

The only difference between Engel's and Buddington's interpretations is one of time.

www.ingramcontent.com/pod-product-compliance
Lightning Source LLC
Chambersburg PA
CBHW030527210326
41597CB00013B/1057